CW01522507

BIRDS
OF
THE INDUS

The Publishers, Oxford University Press,
are grateful to the

EMBASSY OF FINLAND

for their generous support in
making this book possible

BIRDS
OF
THE INDUS

Mubashir Hasan

Photographs by
T. J. Roberts
Mubashir Hasan
Syed Asad Ali
Khan Mohammad
F. J. Koning
Rolf Passburg

OXFORD
UNIVERSITY PRESS

OXFORD

UNIVERSITY PRESS

Great Clarendon Street, Oxford OX2 6DP

Oxford University Press is a department of the University of Oxford.
It furthers the University's objective of excellence in research, scholarship,
and education by publishing worldwide in

Oxford New York

Athens Auckland Bangkok Bogotá Buenos Aires Cape Town
Chennai Dar es Salaam Delhi Florence Hong Kong Istanbul Karachi
Kolkata Kuala Lumpur Madrid Melbourne Mexico City Mumbai Nairobi
Paris São Paulo Shanghai Singapore Taipei Tokyo Toronto Warsaw

with associated companies in Berlin Ibadan

Oxford is a registered trade mark of Oxford University Press
in the UK and in certain other countries

© Oxford University Press 2001

The moral rights of the author have been asserted

First published 2001

All rights reserved. No part of this publication may be reproduced, translated,
stored in a retrieval system, or transmitted, in any form or by any means,
without the prior permission in writing of Oxford University Press.
Enquiries concerning reproduction should be sent to
Oxford University Press at the address below.

This book is sold subject to the condition that it shall not, by way
of trade or otherwise, be lent, re-sold, hired out or otherwise circulated
without the publisher's prior consent in any form of binding or cover
other than that in which it is published and without a similar condition
including this condition being imposed on the subsequent purchaser.

ISBN 0 19 577938 X

Typeset in Times
Printed in Pakistan by
Asian Packages (Pvt) Ltd., Karachi.
Published by
Ameena Saiyid, Oxford University Press
5-Bangalore Town, Sharae Faisal
PO Box 13033, Karachi-75350, Pakistan.

To

Frances and Tom Roberts

*the indefatigable explorers of Pakistan's
fauna and flora*

CONTENTS

CONTENTS

ACKNOWLEDGEMENTS

The idea to present the transparencies in the form of a book was that of my wife Dr Zeenat Hussain, an ardent admirer of fauna and flora. Feeling deeply grieved over the wanton destruction of wildlife in Pakistan she has been ever keen on educating the public on the subject. Her advocacy for a book on birds combined with her gentle encouragement and persuasion for over a decade has made this publication possible.

Discovering the names of all the birds on my transparencies through library research was proving to be a frustrating undertaking. As a result, I was content to be an amateur bird-photographer. But for the time and guidance generously given over a period of years by Dr T.J. Roberts, I would still be ignorant of the names of a great number of species. My indebtedness to him for helping in the identification of slides is deep. He was also generous enough to read the manuscript and make suggestions and comments which have been gratefully acknowledged in the text. Then, he very generously offered some of his valuable transparencies for inclusion in this volume. But more is owed to him. As a person who dedicated his life to the study of fauna and flora, especially avifauna of this region, as a researcher of great energy, integrity and humanity for more than three decades, and as a person with a sense of humility which is the hallmark of a true scholar, Tom Roberts was a source of great inspiration.

I am also deeply indebted to Syed Asad Ali, a keen bird lover, a conservationist and a bird-photographer. He has been most generous in offering some of his best transparencies for inclusion in this volume and has been of great help at all stages of this publication. Also it was through Syed Asad Ali that Mr Rolf Passburg and through Tom Roberts that, Rudi Hess,

Tim Hurrell and Brenda Wheeler generously contributed their valuable transparencies for this volume. This contribution is highly appreciated and gratefully acknowledged.

Special thanks are owed to Mr Khan Mohammad of the Sindh and Punjab Wildlife Departments for generously contributing valuable transparencies to this volume. I cannot adequately express through these lines the debt of gratitude owed to my relatives and friends in Pakistan and India: to Khwaja Shaiq Hasan, his wife and children for their arrangements and company at many many camps where a large number of photographs were taken; to Dr Mehdi Hasan for his unfailing comradeship and assistance at many camps and trips; to Dr Zaki Hasan for his many-faceted supervisory role at camps and trips; to Dr Shubbar Hasan, Sillo and their children for keeping an unceasing flow of films and camera equipment from abroad; to Dr Syeda Saiyidain Hameed for meticulously arranging some very productive photographing trips to Bharatpur and for reading part of the manuscript; to Dr V.A. Pai Panandiker for his kind assistance in arranging researches at Bombay; to Mrs Sarojini Chopra for research in Hindi and Sanskrit names; and last but not the least to Mr Raza Kazim whose generous and enthusiastic assistance and advice in all matters photographic made the production of the transparencies possible. It was he who selected, indeed, purchased for me, the camera and all that goes with it and gave exhaustive instruction on techniques and procedures. He was most generous with his time in reviewing my photographic work and making suggestions for its improvement. I am deeply indebted to him.

I have also to express my sense of appreciation and gratitude to the administration and staff of the Bombay Natural History Society for the permission to use the facilities of the library and museum of the society.

Mubashir Hasan

INTRODUCTION

There was an old owl liv'd in an oak
The more he heard, the less he spoke;
The less he spoke, the more he heard
O, if men were all like that wise bird!

What distinguishes birds from all other living creatures are feathers. Man has two feet as birds have and reptiles lay eggs as birds do, but feathers, no other animal possesses. The quest to discover the origin of birds led scientists to concentrate on the origin of feathers. They came to the conclusion that feathers have developed from the reptilian scale which is remarkably similar to the scales birds have on their legs. Birds, like crocodiles, are now believed to have descended from the first reptiles. The missing link between the reptiles and birds, a flying reptile, was uncovered for the first time in 1861 in the form of a 160 million year old fossil that was named *Archaeopteryx lithographica* (ancient winged creature carved on stone). The discovery settled for the scientists any question regarding the reptilian ancestry of birds (7:594).

The Age of Birds and Mammals began with the ending of the Age of Reptiles, 130 million years ago. Rocks from this period, called Cretaceous, have revealed fossils of birds resembling grebes, cormorants, rails, sandpipers, pelicans, flamingoes and herons. However, it was during the Eocene period which began 65 million years ago that birds acquired their basic forms as fliers, swimmers and runners. According to Welty (6:602), during this period, 80 per cent of the modern orders of birds had made their appearance. Besides the genera of eagles, kites, and vultures, those of ostriches, storks, ducks, pheasants, cranes, bustards, owls, gulls, terns, cuckoos, and woodpeckers had appeared. Fossils from this period indicate that the largest penguin of that time stood as tall as a man and weighed about 120 kg.

Another 30 million years had to pass, that is, during the Oligocene period, before the genera of turkeys, parrots, pigeons, swifts, kingfishers, warblers and sparrows made their appearance. It was a period when great forests covered the Earth. Bird life expanded to its greatest number ever, 150,000 species according to the estimates made by Broadkorb (quoted by Welty 6:605). Then the trend of expansion got reversed.

The last two million years have been a period of repeated glaciation, changing sea levels and great dispersals of all life-forms. Each period of glaciation was followed by a period of warming. The cycle caused extensive extinction of bird life. By the beginning of the recent period, 11,000 years ago, scientists have estimated that only 10,600 species out of 150,000 had survived. The extinction of the species has continued. Fisher and Peterson have estimated that since 1680, eighty-five species of birds have become extinct throughout the world (6:697). Much of the extinction in the recent period is due to man. Spread of agriculture, industry, and human habitation has adversely disturbed the environment for bird life. Its slaughter for food, feathers or sport has been another significant factor. From time immemorial, man has depended upon birds for meat and eggs. Human population explosion, technological advances and cultural changes have led to many permanent changes. The Red Jungle Fowl, *Gallus gallus*, was domesticated in India in about 3200 BC and in China and Egypt in about 1500 BC (1). Today there are over sixty breeds of this fowl. In the United States alone are raised, every year, 2 billion chickens, 100 million turkeys and about 11 million ducks. Also raised for food or sport are geese, pheasants, quails, partridges and pigeons (7:626). As John Keats said:

Thou wast not born for death, immortal Bird!
No hungry generations tread thee down;
The voice I hear this passing night was heard
In ancient days by emperor and clown.

Ode to a Nightingale

Only in recent times, man has begun to realize that without birds the entire balance of nature would change and man's existence on earth would be in jeopardy. Scavenger birds such as vultures, crows and gulls which feed upon carrion, offal and garbage perform a tremendous sanitary service to keep the face of the earth clean and healthy. Birds exercise a tremendous check on the growth of insect life on earth. Insects, their eggs, larvae, pupae or chrysalides constitute a large proportion of their diet. According to one estimate a single pair of tits with their progeny destroy annually at least 120 million insect eggs or 150,000 caterpillars and pupae (3:153). Enormous damage to agricultural crops and produce is prevented by owls, hawks, falcons, kestrels and other birds of prey as these exercise a great check upon the population of rodents. It has been estimated that one pair of house rats having six litters of eight young annually and breeding when three and a half months old without losses or casualties, would, in theory, increase at the end of the year to 880 rats; and at the end of five years to 940,369,152 rats (3:153). Without birds the entire spectrum of the Earth's flora and fauna would change to the great detriment of man and other living creatures.

Birds have been nature's very special gift to our planet. There was a time when man accorded special recognition to birds. History records many instances of man attributing supernatural powers to birds. The Sacred Ibis, *Threskiornis aethiopicus*, was revered by the Ancient Egyptians as the incarnation of Thoth, their moon-god of wisdom and magic. The owl symbolized Athena, the goddess of wisdom in ancient Greece. In the Christian religion, the cock signifies watchfulness and vigilance. The image of the White Stork as the carrier of babies was cultivated among the children of Europe for generations. In the subcontinent, the call of the House Crow from the parapet has long been considered a good omen presaging the arrival of a guest. The Blue Jay is associated with the great Hindu god Shiva. An eagle has often been universally used as a symbol of power and war. The dove is similarly recognized as a symbol of peace and love.

In the valley of the Indus are found birds of the South Asian subcontinent, East Africa, Europe and much of the rest of Asia. According to Roberts, the region serves as the great caravanserai for Eurasian avifauna travellers (2.1:23). Some fly in to stay for the summer or winter while others transit through. Some fly in to breed while others fly out to breed. In the Indus valley can be seen birds of the seas and rivers, of marshes and deserts, of ranges of arid hills and forest-covered mountain systems and of plains intensely cultivated, and plateaux with sparse vegetation. They come from far and wide. The Wilson's Storm Petrel, *Oceanites oceanicus*, which may be sighted off Karachi, breeds on the continent of Antarctica lying about 11,000 km to the south (2.1:29).

There are other birds which fly long distances in north-westerly and north-easterly directions. For example, the White Stork, *Ciconia ciconia*, recovered in southern Pakistan was traced from the population that breeds in Germany. The Lesser Golden Plover, *Pluvialis dominica*, seen around Karachi is known to breed only in the Kamchatka peninsula lying nearly a thousand miles north of Japan (2.1:29).

> I see my way as birds their trackless way,
> I shall arrive! what time, what circuit first,
> I ask not: but unless God send his hail
> Or blinding fireballs, sleet or stifling snow,
> In some time, His good time, I shall arrive:
> He guides me and the bird. In His good time!
>
> Robert Browning in *One Word More*

With the regularity of seasons, birds fly into and out of the region in all directions of the compass. A very large number of species enter the Indus valley from the north and north-west through the valleys of the tributaries of the rivers Indus and Jhelum, namely Kurram, Kabul, Chitral, Swat and Kaghan. Roberts (2.1:26) lists more than 120 species which migrate into the valley after breeding during summer from points in the Palaearctic region, a huge landmass comprising Europe, much of North Africa and all of central and northern Asia. These species include ducks, geese, gulls, waders, rails, raptors, wagtails, buntings, warblers and many others. Most of these migrants breed in central and northern Asia.

The Mediterranean and Saharan adapted species enter the region flying in an easterly direction through southern Balochistan and along the coast of the Arabian Sea (2.1:19). The Pied Crested Cuckoo, *Clamator jacobinus*, flies eastward from East Africa to breed in the plains of the Indus during summer (2.1:442). The Blue-cheeked Bee-eater, *Merpos superciliosus*, spends its winter over a wide range of countries in East Africa. For breeding, it migrates eastward and northward and those of its flights which choose to cross the Arabian Sea arrive in Karachi and the Makran coast from the end of April to the beginning of May (2.1:521). The Kashmir Roller, *Coracias garrulus*, takes a flight path through the Arabian peninsula and the Persian Gulf (2.1:527).

There are two major entry routes into the valley on its eastern side. The first lies in the extreme south-eastern corner of Pakistan and is used by birds flying over or coming from central and southern India. The second route for birds flying in a north-westerly direction lies along the eastern and south-eastern borders of the Punjab. The Spotbill Duck, *Anas poecilorhyncha*, resident of India and countries further east flies into Pakistan in a south-west to north-west direction to breed during summer. Similarly, the Indian Skimmer, *Rynchops albicollis*, flies westward to the waters of the River Indus and its larger tributaries to breed during summer. The Red Turtle Dove, *Streptopelia tranquebarica*, flies westward into the Punjab, Sindh and the NWFP in October and migrates back in March.

The Himalayan species enter the Indus valley in a westerly direction in the areas of northern Pakistan. Some of these turn south-west towards Balochistan to fly further east along the coast towards Arabia while others fly directly north-west from northern Pakistan towards central Asia.

Within the Indus valley great congregations move along the river Indus and its tributaries and it is for a good reason that one of the great flyways of Asia is called the Indus Flyway. Large movements heading towards India also take place across the Punjab. A recent study of the migration of the Common Teal, ringed over a period of years at Bharatpur, India, Ambedkar and Daniels (27:224), states that during outward migration most teals pass through valleys like Kaghan, Kurram, etc. On the basis of the absence of any ring recoveries from Nepal and Sikkim, it appears to the authors that the NWFP of Pakistan is probably the nearest flyway for entry into central Asia before spreading out in the Ob, Yenisey and Lena River Basins in the Siberian part of Russia.

Ornithologists give different figures of the total number of the species of birds in the valley of the Indus and its adjoining lands. In the most exhaustive and authoritative study yet, carried out over a period of more than thirty years, Roberts has concluded that 660 species are known to occur or have recently occurred in Pakistan. This figure includes thirty-two species, reported for the first time by Roberts—a great contribution, indeed, to our knowledge. Previously these thirty-two species were not considered to be occurring in Pakistan.

However, in the ten-volume *Handbook of the Birds of India and Pakistan* by Salim Ali and Ripley, the total number of birds inhabiting Pakistan is much more. The difference arises, firstly, due to the emphasis upon sub-species adopted by Ali and Ripley. For example, Roberts reports the occurrence of three species of Bulbuls; *Pycnonotus leucogenys* (White-cheeked Bulbul), *Pycnonotus cafer* (Red-vented Bulbul) and *Hypsipetes madagascariensis* (Black Bulbul or Grey Bulbul). Ali and Ripley split Roberts' species *Pycnonotus leucogenys* (White-cheeked Bulbul) into *Pycnonotus leucogenys leucogenys* (White-cheeked Bulbul). *Pycnonotus leucogenys leucotis* (White-eared Bulbul), and *Pycnonotus leucogenys humii* (Hume's White-eared Bulbul). Similarly they divide *Pycnonotus cafer* into seven sub-species and report two of these to be occurring in Pakistan. Over the occurrence of *Hypsipetes madagascariensis* (Black Bulbul or Grey Bulbul) the two authorities are of similar view. Furthermore, taking all available evidence into account that a

View from the author's camp of a creek of River Chenab at Trimmu where a number of photographs included in this volume were made.

Red-fronted Serin, Common Rose-finch and Striolated Bunting were photographed at Lalazar, in the Kaghan valley.

bird species has not occurred, is highly unlikely to do so, or has become extinct, Roberts has excluded from his list twenty-seven species which Ali and Ripley included in their list.

The 660 birds of Pakistan reported upon by Roberts may be divided into the following categories on the authority of the maps in his book:

(1) Those which may be found all the year round in a particular area or region. These are called residents and there are 242 such species. A part of the population of 29 resident species emigrates to breed outside as well.

(2) Apart from the residents there are 146 species which breed in Pakistan during summer. Out of these there are 58 species which may be sighted in Pakistan during winter on a regular or irregular basis.

(3) Then there are those which migrate out of the country for breeding during the entire summer season or a part thereof. There are 164 such species. In bird literature these are termed as winter visitors. As a majority of them spend a greater part of the year in Pakistan, it may be more appropriate to call them summer emigrants.

(4) There are 59 species which may be sighted in Pakistan in the course of their spring or autumn passage to or from their breeding grounds.

(5) The balance, and indeed, some among the numbers sighted above, are termed rare, stragglers, vagrants, or rare vagrants.

In 1991, water fowl counts were carried out in Pakistan in 176 wetlands. A total of 1,350,000 water birds of 105 species were counted (24:42). The main concentrations were found in the Chashma reservoir, and at the lakes Haleji, Keenjhar, Hadero and at the Hub Dam. The two endangered species, the White-headed Duck, *Oxyura leucocephala*, and the Ferruginous Duck, *Aythya nyroca*, continue to decline in numbers.

GREBES

Family Podicipedidae

Grebes are water birds. Fossils resembling grebes date back to the Cretaceous epoch of the Mesozoic era, which began 130 million years ago. That makes grebes one of the most primitive of the families of birds (7:601). They eat fish, tadpoles, frogs, water insects, small molluscs, crustacea and vegetation. Their legs are placed far back, specially adapted for swimming and diving. They are eminently suited for a purely aquatic life, never needing to step on terra firma (4:541). Roberts reports the occurrence of five species in Pakistan (2.1:49), whereas Ali and Ripley (1.1:2) mention only three in the entire subcontinent.

Little Grebe or Dabchick
Tachybaptus ruficollis or *Podiceps ruficollis*
Dubdubi or *Pandubi* in Urdu and Hindi
Tubino in Sindhi
Pind in Kashmiri

Dabchick, the smallest of grebes, may be seen around lakes, swamps and flooded depressions over most of Pakistan but rarely in main rivers (2.1:49). During breeding season, i.e., summer, their crown, back and hindneck are dark olive-brown and cheeks and foreneck are a rufous chestnut. Both sexes look alike. They breed in Gilgit and Baltistan but migrate southwards in winter. Breeding has also been reported from Khushdil Khan and Zangi Nawar lakes in Balochistan, Salt Range lakes in the Punjab and

Little Grebe, photograph by Mubashir Hasan

in the last century from east Nara in Sindh (2.1:50). The Dabchick grabs its food under water. It is an excellent diver and underwater swimmer. A smooth dive, without leaving a ripple, far beneath the surface, has earned it its Urdu name *Dubdubi*, one who keeps diving; and *Pandubi*, the one who is an excellent water diver. Anyone who enters water within the vicinity of dabchicks cannot fail to notice that they are not easily disturbed. They will dive to reappear further away or simply patter away a short distance.

Body length 250-290 mm.
Wing-span 400-450 mm.

Red-necked Grebe
Podiceps grisegena

Red-necked Grebe is one and half to two times bigger than Dabchick, the smallest of grebes. Its brownish-black crown extends down to the eye. In winter breeding plumage the foreneck is chestnut and ear tufts are absent (Roberts, 1991, vol. 1, p. 52).

According to Roberts, the grebe is vagrant in Pakistan. It was spotted by him at Lal Soharna Park in Bahawalpur in 1967. It has also been spotted in Rawal and Nimmal lakes as well as in the lakes in the Salt Range. The photograph shown in these pages was taken by the author in Bhraratpur, India, in 1991. It was feeding alone as no other bird was anywhere near. It is reported to feed on fish, tadpoles, frogs, and water insects.

Body length 400-500 mm.
Wing-span 770-850 mm.

Red-necked Grebe, photograph by Mubashir Hasan

——— CORMORANTS ———

Family Phalacrocoracidae

In their typical posture, cormorants sit out on tree stumps, ledges, or high ground with open wings as if waiting to embrace. It is held that cormorants are unable to get water out of their feathers as can most other water birds. Hence they need to dry them up by exposing their body to the sun. This posture may also perform another function, namely, of keeping the body warm as it is sometimes held for a much longer time than required for drying up the feathers (7:65).

Like grebes the cormorants are among the most primitive of the families of birds (7:601). Their main diet is fish. Some species also eat tadpoles, frogs and crustaceans. Roberts reports three species of cormorants in Pakistan. A fourth one, an individual lodged in the Bombay Natural History Society collection, the African Pygmy Cormorant, *Phalacrocorax pygmeus*, was collected in southern Balochistan in 1909. In 1927, it was listed under a wrong name, and was correctly identified in 1965. (2.1:64).

Great Cormorant
**Phalacrocorax carbo sinensis*
Jal-Kawwa or *Pan-Kawwa* in Urdu and Hindi
Wadda Silli in Sindhi
Neiar in Kashmiri

Great Cormorant, photograph by Mubashir Hasan

The Great Cormorant is a water bird and, during winter, is found in all the riverain areas of Pakistan, India and Bangladesh and all the coastal areas. Roberts describes it as abundant in Sindh, and less common elsewhere (2.1:61). Their major population is said to breed in Pakistan but some are migrants who have bred in northern latitudes of Asia.

The Great Cormorant is an excellent swimmer and diver. When not actively pursuing fish, it swims almost submerged in water, with only the longish neck and a thin slice of the back showing (1.1:37). It feeds exclusively on fish which it catches under water. It has been reported to play with a fish as a cat does with a mouse. Ali and Ripley state: 'When satiated, has been observed to dive and catch a catfish about 15 cm long…The bird swam ashore with the fish, dropped it struggling on the bank, picked it up again and carried it back into the water, released it and dived after it, caught it again and brought it ashore, then leisurely repeated the whole performance several times before swallowing it.'(1:38)

In its size the Great Cormorant is like a large duck, and sits upright with wings stretched outwards over long periods during the day. Both the sexes look alike. The adults have greenish glossy body feathers becoming a duller black in winter whilst the throat and body become greenish white (2.1:61).

Body length 800-900 mm.
Wing-span 1300-1500 mm.

Indian Shag, photograph by T.J. Roberts

Indian Shag
Phalacrocorax fuscicollis
Chhota Jal-Kawwa in Urdu and Hindi
Silli in Sindhi

This cormorant is slightly bigger than the Litter Cormorant but slightly smaller than the Great Cormorant, the difference in body size being 150 mm to 200 mm. It is easily distinguishable from other cormorants in breeding season, otherwise it looks the same except in hand. The Shag is found on lakes, swamps, water-filled depressions, and backwaters of canal headworks, singly, in twos or threes, or in great congregations on large bodies of water. Its main habitat extends from the lower reaches of the rivers Chenab and Jhelum to the entire province of Sindh. It is very similar to the other cormorants.

Body length 600-650 mm.
Wing-span 257-276 mm.

Little Cormorant
Phalacrocorax niger
Chhota Jal-Kawwa in Urdu and Hindi
Silli in Sindhi

The Little Cormorant is found on lakes, swamps, water-filled depressions, and backwaters of canal headworks, singly, in twos or threes, or in great congregations on large bodies of water. It may also be seen in rice fields during the late monsoon season (2.1:65). Its main habitat in the valley of the Indus is in Sindh, and to a lesser extent, in the Doabs of the rivers in the Punjab except the northern half of the Sindh-Sagar. It is not found in Balochistan and has been spotted in the Frontier Province only along the Indus river as it enters the plains. During the summer it migrates from the Punjab to Sindh, and its breeding area starts from Sukkur and extends southwards to the Arabian Sea. Both the sexes look alike; both take part in nest building, incubation and care of the young. The Little Cormorant is also found throughout India and Bangladesh.

A little over half the size of the Great Cormorant, the Little Cormorant is glossy black, the throat area varying from a greyish pink in winter to a purple-black in the breeding season (2.1:64), which extends from July to September. Its nesting is also reported as early as the end of May (2.1:64). It swims low in the water, and when alarmed it can submerge itself leaving its head and neck above the water surface. It is mainly a fish eater, but tadpoles, frogs and crustaceans also form part of its diet.

Body length 480-510 mm.
Wing-span 181-205 mm.

Little Cormorant, Indian Shag, and Large Cormorant, photograph by Mubashir Hasan

SNAKE BIRDS

Family Anhingidae

Darter or Snake Bird
Anhinga melanogaster or ***Anhinga rufa***
Panwa in Urdu and Hindi
Silli in Sindhi

The big blackish bird, the Darter, is a member of the *Anhingidae* family. When it is not actively pursuing its prey, only a part of its long snaky neck, its small head, and a straight, sharp bill are visible above the surface. The movement of the head from side

Darter, photograph by T.J. Roberts

to side, up and down, back and forth is snake-like. Hence its name Snake Bird. Its food is almost exclusively fish which it hunts either by catching it in its bill or by impaling the body of its victim with a rapier-like thrust. It is an excellent diver and underwater swimmer, and uses its feet to propel itself (1:44).

The Darter is large but has a slender body and a very long snake-like head and neck. Its upper plumage is black, and the wings are streaked and speckled with silver grey. The head and neck are chocolate brown, the chin and throat are white, while the lower plumage is shiny black. The Darter may be found on all the rivers in the Punjab. It is less common in Sindh, and almost non-existent in the large lakes of lower Sindh (2:66). It is found throughout India, Bangladesh and Sri Lanka. Its breeding season is from June/July to September (1:42).

Body length 850-970 mm.
Wing-span 331-357 mm.

PELICANS

Family
Pelecanidae

Great White or Rosy Pelican
Pelecanus onocrotalus
Hawasil in Urdu and Hindi
Pen in Sindhi

The Great White or Rosy Pelican is the largest bird of the valley of the Indus weighing as much as 12 kg, with a wing-span of 270-360 cms (10.6-11.8 feet). It has an enormous bill 45 cms (over 17 inches) long. A capacious gular pouch of loose skin hangs from the lower mandible and the belief is that a pelican's bill can hold more than its belly. The Rosy Pelican is an all-white bird, tinged with pink, with a tuft of yellowish feathers on its breast. In flight the under-wings show black at the wing-tips contrasting beautifully with the white body of the bird.

There is a North Indian *shikari* (sportsman) saying that if you were to hunt down a pelican, then the fuel-wood of your entire neighbourhood would catch fire! Meaning that in order, properly, to cook the big bird, with its tough meat, you would require all the fuel that you or your neigbours might possess.

A breeding site of the Rosy Pelican in the Great Rann of Kutch, just south-east of Pakistan, was discovered by Salim Ali in 1960 (1:27). It also breeds in Central Asia and along the northern shores of the Black Sea (1:28). Its flight paths of entry into, and return from Pakistan have been traced by Roberts through the upper Kurram valley and across north-western Balochistan in a north-south direction. The number of pelicans

Great White Pelican, photograph by
Syed Asad Ali

in Pakistan is reported to be decreasing at an alarming rate. The wintering areas in Pakistan are confined to the southern parts of lower Sindh and two areas along the coast in Balochistan (1:69).

Pelicans are exclusively fish eaters and use their bill and gular pouch in the form of a scoop to catch fish. They often cooperate to hunt fish by forming a semicircle, or take positions from bank to bank (if the body of the water is narrow enough) to drive a school of fish into the shallows. Bill open, lower mandible trailing in the water, with the head above or submerged, the pouch is used as a landing net for the fish.

Body length 1500-1830 mm.
Wing-span 2700-3600 mm.

Pelicans in flight, photograph by Syed Asad Ali

HERONS, EGRETS, BITTERNS

Family
Ardeidae

Herons, egrets and bitterns are long-legged, wading birds with a long bill and neck, well-suited to feed in shallow water. They are familiar waterside birds all over the subcontinent, where their food of insects, frogs, fish, molluscs, etc. is available. Most of the species may be commonly sighted wading or standing motionless along the banks of lakes, marshes, ponds, rivers, roadside burrow-pits and along irrigated fields—some in open country, others under or along vegetative cover. They have long legs which enable them to wade in shallow water and also long retractable necks to snap at their quarry. They fly with legs outstretched, and neck held retracted well into the body in a flat S shape.

In Urdu and Hindi, all the egrets and a few species of herons and bitterns are called by the generic name of *bagla*. During the rainy season, the flight of egrets with dark grey clouds in the background is a splendid sight of beauty and elegance. 'As white as a *bagla*' is a common Urdu expression. Followed by the word '*bhagat*', meaning a 'disciple' the term '*bagla-bhagat*' is used for a pharisaical person. Egrets, herons and bitterns have earned this reputation on account of their hunting technique. They hunt by freezing themselves motionless and waiting for interminably long periods. As the prey comes within their range, they strike with lightening speed either impaling it or grabbing it in the bill. Tossing it into the correct position, they swallow it head first. Sometimes they hunt by stalking. Neck slightly stretched, out they wade noiselessly with extreme deliberation in as slow a motion as required by the position or movement of the quarry.

Paddy Bird or Pond Heron
Ardeola grayii
Bagla or *Andha Bagla* in Urdu
Broku in Kashmiri

The Pond Heron is one of the most familiar waterside waders found abundantly in the Punjab and Sindh all the year round. It visits the broader valleys of the NWFP only during summer. It is absent from Balochistan except along the coast and the Sibi plain (2:82). It is crow-sized but with longer legs. It must pay a daily visit or two to every accumulation of water in the countryside or on the outskirts of villages and towns. It is

Paddy Bird, photograph by Mubashir Hasan

widespread throughout India, Bangladesh, Sri Lanka and valleys of Nepal.

During non-breeding plumage, its back is ashy brown, head and neck are dark brown with yellowish buff streaks. During flight, its drab appearance is dramatically transformed as its pure white wings and tail flash into view. During breeding plumage, its back turns a deep maroon, head and neck light yellowish brown. Long filamentous plumes cover its back. The pair bond is monogamous. The female builds the nest with material brought by the male. Both share incubation duties and the care of the young (1:83).

Body length 420-450 mm.
Wing-span 750-900 mm.

Cattle Egret or Buff-backed Heron
Bubulcus ibis
Gai Bagla, Badami Bagla, Surkhia Bagla in Urdu and Hindi
Kurk Pakhi in Sindhi

The Cattle Egret derives its name from its habit of keeping company with cattle. It picks off grasshoppers and all types of insects dislodged by a tractor or a pair of bullocks ploughing the land or those dislodged by the feet of moving cattle. It is found throughout the Punjab and Sindh in winter as well as summer. It is found throughout India, Bangladesh, Sri Lanka and the lowlands of Nepal. It is subject to local migration. Studies in Africa revealed that many nestlings ringed at breeding sites were recovered as far away as 3,500 kms (6:76).

The Cattle Egret is an all-white bird, very common, and similar to the Little Egret during non-breeding plumage except for its yellow bill. During breeding period, its head, neck and back turn golden buff and filamentous plumes of the same colour appear on its breast and wings and its bill turns scarlet-red. It wades into water and also feeds on the ground far away from a lake or a river.

Cattle Egret, photograph by Mubashir Hasan

Body length 450-530 mm.
Wing-span 900-960 mm.

Western Reef (Heron) Egret
Egretta gularis
Kala Bagla in Urdu and Hindi

Salim Ali and Ripley describe this heron (egret) as a very common wader on northern shores of the Arabian Sea and seaboard of Pakistan (1:75). Barring the vagrant ones, it is a salt-water bird, occurring in tidal lagoons, mudflats, mangrove swamps and sandy and rocky seashores. It is found along the coastline of Pakistan during summer, and along the river Indus and its vicinity up to 240 kms upstream from the coast (2.1:86).

Western Reef (Heron) Egret, photograph by Khan Mohammad

It is also found on the shores of western India and on the north-western shores of Sri Lanka.

The Reef Heron is polymorphic, a greater portion of its population is slaty grey in colour, while the remaining portion is all white, almost indistinguishable from the Little Egret (2.1:86). During the breeding period, it develops plumes of its own respective colour in the manner of the Little Egret. Its general habits and food requirements are like other herons.

Body length 550-560 mm.
Wing-span 860-1040 mm.

Little Egret
Egretta garzetta
Chhotta Bagla in Urdu and Hindi
Bararo in Sindhi

The Little Egret is a still smaller version of the Large Egret. The plumage is all white and the bill is black. During breeding season, it develops beautiful plumes on the back of its head, shoulders and breast. These plumes, called aigrettes, were much in demand during the early part of this century for fashionable ladies.

The Little Egret may be found on any big or small body of water where its food of fish, frogs or molluscs is available and which is a little away from human approach. It is a resident species of the plains of the Indus and is also found on the big lakes

Little Egret, photograph by Mubashir Hasan

in Balochistan and along the areas of the rivers Kurram and
Swat, but at lower altitudes. It is abundant throughout India,
Bangladesh, Sri Lanka and the lowlands of Nepal. It has the
same general habits and feeding requirements as the other
members of the family *Ardeidae*.

Body length 550-650 cm.
Wing-span 880-950 cm.

Smaller Egret or Intermediate Egret or Median Egret
Egretta intermedia
Patokha Bagla in Urdu and Hindi
Baglo Achho in Sindhi

The Intermediate Egret is smaller than the Large Egret, but it is
also a pure white, long-legged, long necked and tall wader.
Distinguishing between the two in the field is difficult except
during breeding plumage. This egret may sometimes be correctly
recognized through its hunting technique. Unlike the Large Egret
it does not resort to the technique of standing motionless and
waiting for the prey to come within its range before it strikes. It
prefers stalking its prey with continuous cautious movement
(2.1:90).
    Less numerous than the Large Egret, it is found throughout
the Punjab and Sindh along the main rivers and large lakes. It
breeds in several area in Sindh and is found throughout India,
Bangladesh, Sri Lanka and the lowlands of Nepal. It has the
same general habits and feeding requirements as the other
members of the family *Ardeidae*.

Body length 650-720 mm.
Wing-span 1050-1150 mm.

Large Egret or Great White Egret
Ardea alba
Bara or *Malang Bagla* in Urdu and Hindi
Achho Baglo in Sindhi

The Large Egret is a pure white, long-legged, long necked and tall wader standing 750 mm (2.5 feet) high. Its bill is yellow in winter and black in the breeding period. It frequents rivers and large bodies of water and is found all over the Punjab and Sindh and in small numbers at Khushdil Khan lake in Balochistan. It is found throughout India, Bangladesh, Sri Lanka and the lowlands of Nepal.

The Large Egret has the same general habits and feeding requirements as the other members of the family *Ardeidae*. It breeds in Sindh and some areas of the Punjab. Both sexes look alike and share in incubation and feeding the young.

Body length 850-1020 mm.
Wing-span 1400-1700 mm.

Large Egret, photograph by Mubashir Hasan

Grey Heron
Ardea cinerea
Nari or *Sain* in Urdu and Hindi
Saa in Sindhi
Brag in Kashmiri

The Grey Heron derives its name from its general ashy grey appearance. It is a tall bird, standing as high as 750 mm (2 feet 6 inches). It is common all over Pakistan, India, Bangladesh, Nepal and Sri Lanka, frequenting swamps, marshes and edges of lakes and rivers. Its upper plumage is ashy grey; head, neck and under parts are white. Its crest is black and a chain of black streaks in the middle of the foreneck goes down to the breast. Sexes are alike except that the female is a little smaller.

According to Roberts, seventy-five per cent of the population is believed to be migratory. The number that migrates northwards out of Pakistan is not known but their passage has been marked through Gilgit, Chitral and Kurram valleys (2:93). Several breeding sites in Sindh have been recorded.

Grey Heron, photograph by Mubashir Hasan

A watchful and vigilant bird, the Grey Heron senses danger from humans long before certain other water birds and takes to the air with a loud and harsh croak.

Body length 900-980 mm.
Wing-span 1750-1950 mm.

Purple Heron
Ardea purpurea
Lal Nari or *Lal Sain* in Urdu and Hindi

The Purple Heron is a long-legged, long necked, purplish wader found throughout Punjab and Sindh. It is largely absent from Balochistan and NWFP. Both the sexes look alike. It is also found throughout the plains of India, Bangladesh, Sri Lanka

Purple Heron, photograph by Mubashir Hasan

and the lowlands of Nepal. In Punjab and Sindh more than a dozen breeding sites have been spotted. It avoids expanses of open water preferring swamps with reeds which offer excellent camouflage and make its sighting difficult. In size, it is slightly smaller than a Grey Heron. It has a rusty chestnut neck, boldly striped with black, purplish above, its crown and crest are a slaty black. The rest of the head, chin and throat are white.

The Puple Heron is a very watchful and vigilant bird. Like the Grey Heron it senses danger from humans long before other water birds and takes to the air with a loud and harsh croak.

Body length 780-900 mm.
Wing-span 1200-1500 mm.
Standing 700 mm. high

A favourite site of many birds on the waters of River Chenab at Trimmu. Author's camera is also visible.

STORKS

Family Ciconiidae

The members of the family *Ciconiidae* are large to very large, long-legged, long necked, heavily built birds, standing 600 mm to 1200 mm high. They have long stout bills, short tails and long and broad wings. They are strong fliers and fly with legs and neck outstretched. Quick to exploit thermals, they are capable of feats of soaring and some perform long migrations.

Storks find food in open fields or in partly marshy spots although their long legs and partially webbed feet are indicative of their wading capabilities. They are vocally silent as they lack voice muscles. Most nest in trees although some like the White Stork prefer buildings. Both sexes look alike and take part in nest building, incubating the eggs and caring for the young. (11:782)

Painted Stork
Mycteria leucocephala
Janghil or *Dhok* in Urdu and Hindi
Lamjang or *Lungduk* in Sindhi

A unique combination of black, pink, yellow, and orange in bold strokes has earned for this huge white stork the picturesque name of Painted Stork. It is long-legged, long necked and stands 930 mm (3 feet) high (1:93). Both sexes look alike. In Pakistan, it is found mainly along the river and in the major lakes in Sindh and is only occasionally spotted in the Punjab along the

Painted Stork, photograph by Mubashir Hasan

larger river channels in the plains. It is found throughout India, Bangladesh, Nepal Terai and the lowlands of Sri Lanka. In Pakistan, breeding is reported in the Indus Delta (2:97).

Like other storks, it lacks vocal organs, hence no voice or calls. Its population is reported to be rapidly dwindling due to continuous poaching by fishermen who capture and sell its chicks to animal exporters (2:97). Roberts laments that it is now very rare in Pakistan. Its main diet consists of fish, reptiles, frogs and insects.

Body length 950-1050 mm.
Wing-span 1500-1600 mm.

Black-necked Stork
Ephippiorhynchus asiaticus
Loha Sarang in Urdu and Hindi

The Black-necked Stork is a very tall stork, standing 1350 mm (4.5 feet) high. It has a black-and-white body, black bill and

Black-necked Stork, photograph by Mubashir Hasan

very long reddish-orange legs. Over the last thirty years just about a score of sightings have been reported from several places in Sindh and the Punjab. It is widespread but nowhere abundant in India; also found in Bangladesh, Sri Lanka and Nepal Terai. Reports of breeding are scarce. Roberts classifies it as an 'occasional straggler' to Pakistan. Its general habits and feeding requirements are similar to other storks. The photograph included in this volume was taken at the Bharatpur sanctuary in India.

Body length 1520 mm.
Wing-span 2250-2400 mm.

— IBISES, SPOONBILLS —

Family
Threskiornithidae

Threskiornithidae are long-billed marsh birds often seen on the banks of rivers and lakes. Their bills are spoon shaped, long and down-curved and broader in the base. Both the sexes look alike. They normally associate with and are related to storks, herons, and egrets and are found almost throughout the world: North America, South America, Europe, Australia and Africa besides Asia. Some of its species are migratory. They frequently associate with grazing buffaloes, feeding on insects, snails, fish, etc, often wading in shallow water and submerging their whole head and neck.

White Ibis
Threskiornis melanocephalus, also
Threskiornis aethiopica
Safed Baza in Urdu and Hindi

The ancient Egyptians considered this Ibis to be the incarnation of Troth, their god of wisdom and learning, hence the name Sacred Ibis for the species *Threskiornis aethiopica*. It is a very long-legged, all white bird, larger than a kite but smaller than a vulture. It has a long black down-curved bill and a naked black head. Both sexes look alike.

The Ibis feeds on fish, frogs, molluscs, insects, worms and some vegetable matter. Being a wader, the Ibis is found in marshlands, along rivers, *jheels*, seepage zones; also in

White Ibis, photograph by Mubashir Hasan

mangroves and salt water marshes. It is found all over India. In Pakistan, it is scarce and is confined only to the far south of Sindh. Only one record of sighting exists in the Punjab, at Marala barrage. It is described as a resident of the Indus Delta, but there is no recent breeding record available.

Body length 650-750 mm.
Wing-span 1120-1240 mm.

Spoonbill
Platalea leucorodia
Chamcha or *Chamcha Baza* in Urdu and Hindi

The Spoonbill is an all white bird, standing 600 mm (2 feet) high. It is as big as a large duck, but with a much longer neck and legs. Its bill is shaped like a spoon, hence the bird's name Spoonbill in English and *Chamcha* in Urdu. In breeding plumage, the tip of the bill becomes yellow and a crest of pale yellow feathers extends down the back of the neck with a yellow

Spoonbill, photograph by Mubashir Hasan

patch on the upper breast. It feeds in shallow water, the partly open bill sweeping the bottom in a semicircular motion, sifting small food items comprising fish, tadpoles, frogs, molluscs, crustaceans, aquatic insects and some vegetable matter.

The Spoonbill is found all over Pakistan and India feeding or resting on the edges of large lakes and rivers. However, due to ruthless hunting practices, it is becoming scarcer in Pakistan. Two-thirds of the population is considered to be migratory, entering through Balochistan, the Kurram valley and the Indus (2:112). Ringed birds from Turkey and the former USSR have also been recovered in Pakistan. Breeding activity has been cited from southern Sindh and Nara areas.

Body length 850-950 mm.
Wing-span 1200-1350 mm.

━━ FLAMINGOES ━━

Family
Phoenicopteridae

This is the family of flamingoes. The specie is known for its very long legs, wry slender necks and pinkish plummage. When sighted in large numbers, they give the impression of a huge bed of pink and white roses. The females are usually smaller than the males in height. Colonial in nesting and gregarious in feeding, sometimes remaining for periods of several months in one feeding area, found around large inland lakes or along the sea coast or on muddy banks. They are quite shy and will take to flight if a human approaches within 200 metres. They feed on micro-organisms, shrimps, algae, small insects and also on seeds of rice, etc.

Greater Flamingo
Phoenicopterus ruber (2:113) **or**
Phoenicopterus roseus (1:118)
Raj Hans in Urdu and Hindi
Lakka or *Lakke Jani* in Sindhi

The tallest bird of the Indus Valley, the Flamingo may stand as high as 1550 mm (5 feet 1 inch). It is found regularly in the brackish water lakes of the Punjab Salt Range, in the large lakes of Sindh and along the entire coast of Balochistan and Sindh. Ali and Ripley report its sporadic and capricious occurrence throughout India. It may be sighted throughout the year but it is considered to be a largely migratory bird although not much is

Greater Flamingo, photograph by Mubashir Hasan

known about the migratory pattern. It is reported to breed in Afghanistan, Iran and in the Great Rann of Kutch, a part of which lies within the borders of Pakistan.

The Flamingo has a rosy white body, striking red and black wings, and a very long neck and legs. The female is slightly smaller in size. It feeds on crustaceans, water insects, and molluscs by a specialized technique of filtration of mud and water pumped in by the tongue with the upper side of the bill, scraping the bottom.

Body length 1250-1450 mm.
Wing-span 1400-1650 mm.

Khubakki salt water lake in the Salt Range where the Flamingoes, Cinerous Vulture, Imperial Eagle, and Great Plover were photographed.

– DUCKS, GEESE, SWANS –

Family
Anatidae

The members of the *Anatidae* family are water birds with longish necks and narrow and usually pointed wings. Some migrate over long distances. The flight is strong and direct with a rapid beat of wings and is relatively fast. Many species fly in large flocks. In all there are 247 forms, belonging to 147 generally accepted species, 43 genera and 10 tribes (11). Only the Mallard and Muscovy duck of South America have been domesticated.

All ducks, swans and most geese moult the flight feathers of the wings simultaneously and pass through a flightless period from three to four weeks after the breeding season. They are excellent swimmers. They have short legs and their feet are webbed. Many species obtain their food by diving. Their bill is broad, flat, rounded at the tip and has a comb-like fringe for straining out food particles from water. Most feed on aquatic plants and animals while some graze for vegetable food on land (12).

Greylag Goose
Anser anser
Qaz, *Kaj* or *Hans* in Urdu and Hindi
Hanj in Sindhi
Mug in Punjabi

The Greylag Goose is very similar to the domesticated goose except for its pink bill and grey-brown breast plumage. Both

Greylag Goose, photograph by Mubashir Hasan

sexes look alike. It is found in the riverain areas of the Chenab, Jhelum and the Indus, as well as in the major lakes of Pakistan from October/ November to March but is scarce. It is also found throughout northern India, Nepal and Bangladesh. During March it begins to migrate to its breeding grounds in Central Asia—from the former USSR to northern China. The migration route lies through the high northern routes from Baltistan to Gilgit and north-western routes of Balochistan and of the valleys of the rivers Kurram and Kabul. The return migration takes place in October/November.

The goose is an extremely alert bird, very shy of humans. During the day, gaggles of geese roost in main river channels and on sandbars, which are difficult to approach by humans—an individual or two are always on watch duty as others sleep. It feeds at night in big and small *jheels*; occasionally on shoots of wheat and gram crops; exclusively vegetarian. A gaggle of geese can be very noisy but sometimes huge formations can fly past at a stretch without uttering a sound. It flies in V-formation and the flight is fast and direct.

Body length 800-900 mm.
Wing-span 1520-1800 mm.

Bar-headed Goose
Anser indicus
Qaz, *Kaj*, or *Hans* in Urdu and Hindi

The Bar-headed Goose is a little smaller than the domesticated goose. A black bar from eye to eye at the back of the head and another just below it is the distinctive mark of this otherwise grey, brown and white bird. Both sexes look alike.

It is rare in Sindh and scarce in the riverain areas of the Ravi, Chenab, Jhelum and in the reaches of the Indus in the Punjab. It is also found throughout northern India, Nepal and Bangladesh, and is rare in southern India. It arrives in Pakistan in October/ November and is well settled for wintering by mid-December.

Both the Bar-headed Goose and Brahminy Duck can be seen here, photograph by Mubashir Hasan

By the end of March it has left for its breeding grounds in Central Asia, Tibet and Ladakh.

Like the Greylag goose, the Bar-headed is an extremely alert bird, very shy of humans. During the day, gaggles of geese roost in main river channels and on sandbars, which are difficult to approach by humans—an individual or two always on watch duty as others sleep. It feeds at night in big and small *jheels*; occasionally on shoots of wheat and gram crops. It is strictly vegetarian. Like the Greylag, the Bar-headed can be very noisy but if the situation demands, sometimes huge formations can observe complete silence. It flies in V-formation and the flight is fast and direct.

Body length 700-780 mm.
Wing-span 1500-1650 mm.

Ruddy Shelduck or Brahminy Duck
Tadorna ferruginea
Surkhab, *Chakwa* (male), *Chakwi* (female) or *Lal* in
Urdu and Hindi
Mungh, *Lalo*, *Kwancha* or *Khatiun* in Sindhi

The Brahminy is a large-sized duck which in Urdu and Hindi folklore is a dream bird. One way of saying that a person is not 'special' is the caustic remark that the person 'is not wearing a feather of *Surkhab*', meaning that he or she lacks the touch of a Brahminy Duck, beauty and fidelity. A man and a woman deeply in love are also called *Chakwa-Chakwi*. The legend goes that the frequent desperate and loud calls of a Ruddy Shelduck pair are indicative of their love for one another. Pair bonds are reported to be monogamous and probably life-long (2:128).

During winter, it is found all over Pakistan, northern India, Nepal and Bangladesh, in major rivers, large lakes, backwaters of canal headworks and along the coast. By mid-April it departs for its breeding grounds in central Asia, southern China, Iran, Afghanistan and Ladakh. The return migration takes place in

October/November. Its diet consists of molluscs, aquatic insects, shoots, tubers, reptiles and grain. Feeding is mainly at night. During the day it rests on sandbars which are practically inaccessible. Very alert and wary of human approach, it is prompt in giving an alarm call of approaching danger.

Beautifully coloured, the rich orange-brown Brahminy is a large duck with a black tail and glossy metallic green and white on black wings. In flight, it presents a magnificent view of an orange-brown body, white underside, and black quills.

Body length 610-670 mm.
Wing-span 1210-1450 mm.

Cotton Teal, Quacky-Duck or Cotton Pygmy Goose
Nettapus coromandelianus
Girri, *Girria* or *Girja* in Urdu and Hindi
Baher or *Kararhi* in Sindhi

A very small goose-like duck, smaller than the Common Teal (2:129). In breeding plumage (mid-March to mid-November),

Cotton Teal, photograph by Mubashir Hasan

the male has face, neck and underparts white and a shiny green-black crown, the back is black glossed green (2:129). A set of black half-collars at the base of the neck leave the centre white. The female is duller and browner with less contrasting white in plumage. In eclipse, the male is like the female. It is called Quacky Duck because of the rapid quacking of the male 'which has almost human tonal qualities' (2:130).

The Cotton Teal is a subtropical bird which breeds in Pakistan, India, Bangladesh, China, South-East Asia, Indonesia and north-east Australia. Earlier, pre-1947, it was believed to be rare; since then, Roberts (2:129) has produced enough evidence of its resident status in Sindh and at the canal headworks at Kalabagh and year-round irregular visitor status along the Indus river channel. It feeds on seeds of water plants, young shoots, rice grains and fresh water crustacea, insect larvae and worms.

This teal is a perching duck nesting in natural hollows in tree-trunks standing near water. Ali and Ripley (1:191) mention 'One nest recorded in a box-like hole in the coping of Government House, Rangoon, in 1924, 68 feet (c. 20 m) above ground...thirteen ducklings were pushed out of nest-hole by parent, dropped like stones for some distance, then fluttered to break fall, and reached ground safely. Elsewhere female has also been observed carrying down duckling on her back (H.S.Wood).'

The Cotton Teal is a small duck with a short stubby goose-like bill. The face, neck and the lower plumage is cotton white. Both sexes have a generally uniform brown upper plumage, the female less so. The undertail coverts and vent are white. The wings have a prominent white bar. In breeding plumage the male has a largely white face, neck and belly. The back, wing shoulders and tail are black, glossed with green. The crown is shiny iridescent green-black not reaching down to the eye.

Body length 310-340 mm.
Wing-span 500-600 mm.

Nakta or Comb Duck
Sarkidiornis melanotos
Nakta in Urdu and Hindi
Karo Hanj in Sindhi

The Nakta is a very large goose-like, vegetarian, tree-perching duck, glossy black above and white below with a spotted head and neck. The males are as large as the Bar-headed Goose, the females are Mallard-sized. During breeding season, the male develops a fin-like projection on its bill. In the winter it shrinks and becomes more of a rounded knob. The Nakta frequents well-wooded countryside and fresh water ponds. It is found all over India. It was last reported to be breeding only in Thatta district and south-eastern Sindh. Roberts considers its status in Pakistan today as 'probably extinct' (2.1:131). The photograph shown was taken in Bharatpur, India.

Body length 700-780 mm. (males) 500-600 mm. (females)
Wing-span 1400-1600 mm. (males).

Nakta, photograph by Mubashir Hasan

Wigeon
Anas penelope
Peason or *Chhota Lalsar* in Urdu and Hindi
Pharao in Sindhi

The Wigeon with a reddish-brown head is slightly smaller than the Mallard. It is a palaearctic migrant of Eurasian distribution, breeding in its northern part and wintering in the south from Portugal to China and Japan. Its northward migration takes place in March and southward in October. It is abundant throughout South Asia frequenting lakes and reservoirs which have grassy edges. It is a non-diving duck, and feeds while swimming or walking about by grazing or grubbing. It is mainly vegetarian, and feeds in rice fields as well.

In the breeding season the head and neck of the male are rusty red and the body silvery grey. In the non-breeding season the upper plumage is reddish-brown marked and vermiculated with black and the lower plumage is chiefly white. The female is similar to the male but less reddish-brown and it has two colour phases; one redder, the other greyer (1.1:168).

Body length 450-510 mm.
Wing-span 750-860 mm.

Wigeon, photograph by Mubashir Hasan

Gadwall, photograph by Mubashir Hasan

Gadwall
Anas strepera
Myla, Beykhur or *Bhuar* in Urdu and Hindi
Buhar, Buari or *Burd* in Sindhi

The Gadwall is a little smaller than the Mallard and is the least brightly plumaged among dabbling ducks. It is totally vegetarian and is largely confined to lakes and swamps and is quite common in the Punjab and Sindh. It is found throughout Bangladesh, Nepal and in the northern half of India.

The female Gadwall is a little smaller but much like the female Mallard, dark brown and buff, streaked and spotted with black. The male is similar to the female but greyer and with a chestnut patch on the wings in breeding season.

By mid-March, the Gadwall leaves for its breeding grounds in the far north. One individual ringed at Manchar lake was recovered 3,000 kms away in the Omsk region in Siberia. The return migration starts in Ocober. It breeds in the central latitudes of Eurasia from China to Western Europe but a few also winter in Balochistan (2:133).

Body length 460-560 mm.
Wing-span 840-950 mm.

Common Teal
Anas crecca
Chhoti Murghabi or *Teal* in Urdu and Hindi
Kardo in Sindhi

The Common Teal is the most abundant duck throughout south
Asia. All that is required is a big or small stretch of water or a
marsh, not much disturbed by humans, and not too deep but
with vegetation and it will attract flocks of teal. By April it has
left for its breeding grounds which are spread over northern
Asia and Europe. It begins to arrive back in August. The analysis
of the ringing data, by Ambedkar, suggests that for the birds
ringed at Bharatpur, India, 'the NWFP of Pakistan is probably
the nearest flyway for the entry to the Russian territory before
spreading out in the Ob, Yenisey and Lena River basins in the
Siberian part of USSR' (Journal, *BNHS* Vol. 87, No 3. p. 417).

The Common Teal is a small duck, nearly two-thirds the size
of larger ducks like Mallard or Pintail. The male has a chestnut
head, a broad metallic green band with a narrow cream border,
running from the eye downward towards the nape and a

Common Teal, photograph by Mubashir Hasan

tricoloured wing-bar—black, green and buff. The female lacks the multicoloured head; is mottled dark and light brown and has a speckled brown throat.

The Teal is an expert flier, extremely fast with an amazing ability for sudden changes of course in all directions, especially vertically upwards. It is almost entirely vegetarian.

Body length 340-380 mm.
Wing-span 580-640 mm.

Mallard
Anas platyrhynchos
Neelsir in Urdu and Hindi
Niragi or *Hiragi* (male), *Niragiani* (female) in Sindhi

The drake Mallard is the prince of ducks, a beautifully shining bird, with a metallic dark green head and neck, maroon colour of the breast separated from the green of the neck by a white collar. Its body grey, tail white, rump black, belly white; black central tail feathers curled up and forward were sought by dandies of yesteryear to decorate headgear. The female looks quite different from the male; brown to dark brown, heavily streaked and spotted, belly grey-brown. It is the ancestor of all domesticated ducks.

The Mallard is concentrated on large bodies of water during the day. Flies out to feed at night mainly upon submerged vegetation, seeds and paddy; larvae, insects, molluscs and water beetles are also consumed. It is a strong and fast flier, can rise vertically upwards from the water surface and cruise at 80 kms per hour. Females quack loudly, the male calls softly and less frequently, a guttural 'roehb' (Roberts personal comments).

It is found from November to the middle of March in somewhat patchy distribution, throughout Pakistan, Bangladesh, Nepal and the northern half of India. It migrates towards its

Mallard, photograph by Mubashir Hasan

breeding in northern Asia, Kashmir, and Tibet. It also breeds in Europe-Turkey, northern Iran and Afghanistan.

Body length 500-650 mm.
Wing-span 810-980 mm.

Spotbill Duck
Anas poecilorhyncha
Hanzar or *Gugral* in Urdu and Hindi
Hanjar in Sindhi

The Spotbill is as large as a Mallard or Pintail, it has grey and dark brown plumage with a spotted breast and scalloped flanks. Two orange-red spots at the base of the bill give this duck its name. It frequents vegetation covered waters and is rarely seen in open rivers and is found throughoout South Asia. It was

Spotbill Duck, photograph by Mubashir Hasan

reported to breed in India and Bangladesh only and was considered a monsoon visitor to Pakistan. Roberts reports breeding activity in the lower reaches of the Indus, Ravi and Sutlej, as also at Kushdil Khan lake in Balochistan (2:37). It is reported to be present all the year round in the Indus Delta, Balloki, Sidhnai and the Nara area. Its food is mainly vegetarian: young shoots, weeds, rice, etc.

Body length 500-650 mm.
Wing-span 810-960 mm.

Pintail
Anas acuta
Seekhpar or Sand in Urdu and Hindi
Drigosh in Sindhi

The Pintail is a large duck frequenting lakes, backwaters of headworks and rivers from October to March all over Pakistan, India, Bangladesh, Nepal and Sri Lanka. It is a very handsome duck, richly chocolate, white, black and grey in colour with a

Pintail, photograph by Mubashir Hasan

long spiky tail, the latter giving it its Urdu name *Seekhpar*—the one with spiky feathers.

The flight of the Pintail is fast and produces a swishing sound audible from considerable heights. It is a very alert duck, being the first to take-off at the slightest suspicion of the approach of hunters. It will also avoid flying within gunshot range of anyone looking like a hunter. Its flocks tend to separate into all-male and all-female gatherings (2:39). It breeds in northern Europe and Central Asia right up to the eastern-most areas of the former USSR. Food mainly consists of vegetable matter, but worms, aquatic insects, etc. are also consumed.

Body length 510-660 mm.
Wing-span 800-950 mm.

Garganey
Anas querquedula
Chaita, *Khira* or *Patari* in Urdu and Hindi
Charho, *Kardo* or *Kararo* in Sindhi

The Garganey is also called the Garganey Teal or Bluewinged Teal. It is one of the commonest and most widespread of the ducks and may be found on every type of water large or small, feeding on seeds and tender shoots of marsh plants and grasses and grains of rice. One of the last of the ducks to leave for its breeding grounds in Central Asia right from China to Western Europe. It is gone by the end of April and is back by August being one of the last of the ducks to leave and the first to return. In Pakistan it is most abundant during the two migration periods to and from India and Sri Lanka.

The Garganey is slightly larger than the Common Teal and much smaller than the Mallard (2:139). The male in breeding plumage has a chocolate brown head and neck with a white stripe through the eye. The wing shoulders are bluish-grey and the lower plumage is white. The female is much like the Common Teal but has more white on the belly. In eclipse

Garganey and Shoveler, photograph by Syed Asad Ali

plumage the male is like the female but the wings remain coloured.

Body length 370-410 mm.
Wing-span 600-630 mm.

Shoveler
Anas clypeata
Tidari, Punana, Tokarwala, or *Ghirah* in Urdu and Hindi
Alipat, Gaino, or *Langho* in Sindhi

The Shoveler is so named because it has a broad shovel-shaped bill. It is as big as the Mallard. It is the third most abundant duck of the valley of the Indus (2:141) and is found in all the main rivers and large bodies of water. It leaves for its breeding grounds in the north by April and begins its return in August. It breeds in Siberia and northern areas of Central Asia and northern Europe.

Shoveler, photograph by Mubashir Hasan

The Shoveler usually keeps in parties and feeds while swimming slowly with neck and broad shovel bill stretched stiffly in front sifting out from the water by the comb-teeth fringing the bill. Its food consists of crustaceans, molluscs, water insects and larvae, fish spawn, worms, etc. with a quantity of vegetable matter, shoots and aquatic weeds (1:173).

The Shoveler is a strikingly coloured duck; the head and neck of the male are glossy metallic green, the breast is white and the lower plumage is reddish chestnut except for a white patch separating black tail coverts. The female is mottled dark brown and buff like the Mallard but with a greyish blue shoulder patch.

Body length 440-520 mm.
Wing-span 700-840 mm.

Red-crested Pochard
Netta rufina
Lal Chonch or *Lalsir* in Urdu and Hindi
Batsha or *Rutabo* in Sindhi

The Red-crested Pochard is most handsome but is not a very common duck. It may be found on the large bodies of water of the canal headworks on the Indus and on some of the larger and deeper lakes with underwater vegetation and during southward and northward migration, along the Indus and the Jhelum. By mid-March it leaves for its breeding grounds in Central Asia and returns in October. It is also found in Nepal and throughout northern India and to a limited extent in the south. It is largely vegetarian and feeds by diving.

The male in breeding plumage has a silky orange chestnut head, bill bright red, neck, breast and tail coverts black, flanks and wings white. Females have a dark brown crown, paler neck and cheeks, above dull brown, below largely white except the breast. The bill is bluish with a pale tip.

Body length 530-570 mm.
Wing-span 840-880 mm.

Red-crested Pochard, photograph by Mubashir Hasan

Common Pochard, photograph by Mubashir Hasan

Common Pochard
Aythya ferina
Burar Nar or *Lalsir* in Urdu and Hindi
Torando in Sindhi

The Common Pochard is abundant in the valley of the Indus on rivers, headworks, ponds, large lakes where water is open and deep and food can be had by diving. In India it is widespread in the north-western areas but not so much in the east or south. It leaves for its breeding grounds in Central Asia and northern parts of Europe by the end of March and is back by late October or early November. It is largely vegetarian, feeds on buds, shoots, seeds; also crustacea, water insects, larvae, etc. feeding both during daytime and at night.

The Commmon Pochard is short-necked and a little smaller than the Mallard. The head and neck of the male in breeding plumage is chestnut red, breast black, and the upper plumage light grey. The lower plumage and sides largely greyish. The head, neck, breast and upper back of the female are rufous brown; the back greyish brown and the bill is blue-grey with black at the tip and base.

Body length 420-490 mm.
Wing-span 720-820 mm.

Ferruginous Duck or White-eyed Pochard
Aythya nyroca
Kurchiya or *Burar Mada* in Urdu and Hindi
Burnu or *Burino* in Sindhi

The Ferruginous Duck is not very common and is found only in large lakes, backwaters of canal headworks and main rivers where reed cover is plentiful—mostly in Sindh and south-western Punjab. It breeds in Iran, Afghanistan, Kashmir, Ladakh, Tibet, southern and eastern Europe and Turkey but evidence of breeding in Zangi Nawar lake in Balochistan is also on record (2:147).

Ferruginous Duck, photograph by Mubashir Hasan

The Ferruginous Duck is an expert swimmer and diver and feeds on vegetable and animal matter such as seeds of water plants, shoots and leaves; and crustaceans, molluscs, water insects and worms.

The Ferruginous is a brown duck, a little smaller than the Mallard. The male is dark maroon chestnut on head and flanks with conspicuous white undertail coverts. The back and wing coverts are darker brown. The female is similar to the male but duller brown. The iris of the male is white and that of the female brown.

Body length 380-420 mm.
Wing-span 630-670 mm.

Tufted Duck
Aythya fuligula
Ablak, *Dubaru* or *Rahwara* in Urdu and Hindi
Turando or *Runharo* in Sindhi

The Tufted Duck is a little smaller than the Mallard and is found in large numbers at canal headworks and large lakes. It begins to leave in March and arrives back in September. On passage it may be found in all the major riverain areas of the Indus and its tributaries. It is also common in Nepal and northern India but less so in the south and Bangladesh. By April it leaves for its breeeding grounds. It breeds in northern latitudes, from the United Kingdom right across northern Europe and the former USSR upto northern China and Korea.

The Tufted is a diving duck which feeds on vegetable and animal matter such as seeds of water plants, shoots and leaves; and crustaceans, molluscs, water insects and worms. The plumage of the Tufted Duck is black-and-white, hence the Urdu name *Ablak*; the head, neck, breast, and tail are black, and the rump is white. The drake has a drooping black crest not found in any other duck (5:58).

Body length 400-470 mm.
Wing-span 670-730 mm.

Tufted Duck, photograph by Khan Mohammad

EAGLES, HAWKS, KITES, BUZZARDS, HARRIERS, ─── VULTURES ───

Family
Accipitridae

The diurnal birds of prey, that is those which hunt or look for food during the day are members of two families, *Accipitridae* and *Falconidae*. They comprise 220 species in over 60 genera (15:5). According to Stresemann (quoted 1:338), *Falconidae* moult the primaries in a distinct and different sequence and this distinguishes them from all *Accipitridae*. The birds of prey have short powerful bills, hooked at the tip, suited to piercing and tearing flesh. Most are capable of killing prey equal to half their own weight but some can do better (15:9). They have strong legs and hooked powerful claws. Both sexes are nearly alike, the female slightly larger. They breed in trees or in crags.

According to Salim Ali and Ripley, birds of prey are 'difficult to identify in the field, and often even in the hand...There has been a great deal of confusion in their identification and distribution in the past' (1.1:282).

Honey Buzzard, photograph by Syed Asad Ali

Honey Buzzard or Crested Honey Buzzard
Pernis apivorus or *Pernis ptilorhyncus*
Shahutela in Urdu and Hindi

The Honey Buzzard is found all over central Punjab and Sindh; also throughout India and in Nepal. The population in Sindh increases in winter and in the Punjab in the spring.

Since larvae of *Hymenpoptera* and beeswax is the chief item of Honey Buzzard's food, its migratory pattern corresponds with that of the bees. It arrives in the Punjab at the same time as the migrating Rock Bee (2:158). Along with honey it consumes larvae of bees. Its head, body and legs are well-protected against the onslaught of the most ferocious of bees. Examination of stomach contents has revealed quantities of wax also. Large insects, reptiles, mice and young birds are also its prey. It breeds in summer in the Punjab and to a lesser extent in Sindh.

Roberts considers that the Crested Honey Buzzard occurring in north-west India and Pakistan is similar to the European Honey Buzzard (*Pernis apivorus*). In size, it is as big as the Pariah Kite and in colouration, quite variable. In its commonest phase, it is greyish brown above and pale brown below, cross-barred with white; the head is darker grey. In Pakistan, darker forms are less commonly observed than paler grey or white breasted forms, though all colour phases have been noted. (2:158]). Both sexes look alike.

Body length 520-680 mm.
Wing-span 1350-1500 mm.

Black-shouldered Kite
Elanus caeruleus
Kapassi in Urdu and Hindi

The Black-shouldered Kite is crow-sized and is found all over the Punjab and Sindh and parts of the Frontier and Balochistan and east from the Himalayan foothills to southern India. Often seen perching on a pole or electricity line or a tree top, surveying for lizards, rats, mice, snakes, frogs, grasshoppers, crickets, young and sickly birds.

This kite looks for food also by flying over a selected area with sluggish flight, but in striking range rushes upon the quarry,

Black-shouldered Kite, photograph by Mubashir Hasan

taking it away in its claws. The nesting season varies, in Sindh, April to July (2:160). Normally the male brings in the food, and the female feeds the young.

The Black-shouldered Kite is a striking grey and white bird. Its head, neck, underparts and tail are white; crown, back of the head and upper tail coverts are pale grey.

Total length 310-350 mm.
Wing-span 750-870 mm.

Pariah Kite, Indian or Pakistani Kite
Milvus migrans
Cheel in Urdu and Hindi
Ill in Punjabi
Siriun in Sindhi

The Pariah Kite is common in towns and villages of Pakistan, India, Bangladesh, Nepal and Sri Lanka. It is resident all over

Pariah Kite, photograph by Mubashir Hasan

Pakistan except the northern and some north-western areas where it is a summer visitor.

The Pariah Kite is an expert flier and a pirate. In a street crowded with people, with amazing suddenness, it can swoop down, turning and twisting, avoiding all obstacles such as electric wires or moving vehicles, to scoop up a morsel lying on the road. There would hardly be a child carrying meat purchased from a shop who has not had the experience of the Pariah snatching out of his hands a piece of meat he had just purchased to take home.

The Pariah Kite greatly resembles two other sub-species which occur in Pakistan, namely: the Black Kite (*Milvus migrans migrans*) has a whitish head and neck against the fulvous brown of the Pariah, and the Eared or Large Indian Kite (*Milvus migrans lineatus*) is a little larger in size. The Pariah is a large, somewhat darker fulvous brown hawk. It is the commonest and the most abundant of the three sub-species. The forked tail of its specie sets it apart from all other birds of prey.

Body length 550-600 mm.
Wing-span 1600-1800 mm.

Brahminy Kite
Haliastur indus
Brahminy Cheel, *Roo Mubarak* or *Dhobia Cheel* in
Urdu and Hindi
Pilyo or *Rutta Okab* in Sindhi

As big in size as the Pariah Kite, it derives one of its Urdu names, *Dhobia Cheel*, literally washerman's kite, from its white head, neck and breast, contrasting with the rest of the body which is rusty red or deep chestnut and perhaps also because like the washermen in the subcontinent, it is never far from bodies of water. It is common in Sindh, rare in the Punjab and has not been reported from Balochistan and the Frontier. It is found throughout India, Bangladesh, Nepal and Sri Lanka.

Brahminy Kite, photograph by Khan Mohammad

The Brahminy feeds on fish, frogs, lizards, mud-skippers, etc. It has been frequently observed to flop down on the water in an attempt to pick up small fish off the suface, and to take off again without effort (1.1:231).

Body length 480 mm.
Wing-span 1500-1700 mm.

Ring-tailed Fish Eagle or Pallas' Fish Eagle
Haliaeetus leucoryphus
Machharang, *Machhmanga*, *Dhenk* or *Patras* in Urdu and Hindi
Kural Baaz in Sindhi

Roberts reports that the Ring-tailed Fish Eagle is becoming 'increasingly rare' in Pakistan. In 1974, its population was estimated to be less than forty pairs. It is found mainly along the Indus, south of Mianwali, where it is resident. It is also found in Nepal's lowlands, northern India and Bangladesh.

It is a fish-eater, and will not hesitate to strike a fish as heavy as 6 kg and drag it along the surface to the water's edge (1.1:290). It does not dive in water to catch fish but picks up

Ring-tailed Fish Eagle, photograph by Mubashir Hasan

individuals it can lay its powerful talons upon near the surface. It is also a determined hunter of waterfowl and catches them by surprise or by tiring them out through constant swoops. It also pirates upon other birds of prey and snatches their catches.

As big as the White-backed Vulture, the Ring-tailed Fish Eagle has a dark brown body and pale brown head and neck. Both sexes look alike, though the male is slightly smaller than the female.

Body length 760-840 mm.
Wing-span 2000-2500 mm.

Himalayan Bearded Vulture or Lammergeier
Gypaetus barbatus
Argul in Urdu and Hindi

The Bearded Vulture is found in Baltistan, Gilgit, Kohistan, Chitral and southward along the Pakistan-Afghanistan border down to Quetta, as well as in northern Sindh and southern Frontier province. In India it is purely a mountain bird found in

Himalayan Bearded Vulture, photograph by Rudi Hess

the Himalayas up to Bhutan in the east. It is a hardy bird capable of withstanding sub-zero temperatures.

Like other vultures, it feeds on carrion but also consumes bones. It breaks large bones by dropping them over rocks from a height of fifty to seventy metres and then swallows the fragmented pieces.

It is a very large vulture with neck jutting out in flight more like an eagle than a vulture. Its head and neck are whitish, the body is greyish near the neck, turning into creamy white towards the tail. Below it is pale rusty or creamy white. Both sexes are alike and both have a 'beard'—that is, black bristles extending from the eyes to the base of the bill.

Body length 1000-1150 mm.
Wing-span 2660-2820 mm.

Egyptian Vulture or Scavenger Vulture
Neophron percnopterus
Safed Gidh or *Safed Cheel* in Urdu and Hindi
Hil in Sindhi

The Egyptian Vulture is locally common in Pakistan except the western-most parts of Balochistan. It is found throughout India except the eastern-most parts. It feeds on carrion, human excreta, offal and garbage and is a very useful bird in keeping the environment clean.

Slightly larger than the Pariah Kite, the adult Egyptian Vulture has a whitish body, pale head and neck and black tipped wings. The plumage of the young is dark brown which becomes greyish before it turns whitish in four to five years. Both sexes look alike.

Body length 600-700 mm.
Wing-span 1550-1800 mm.

Egyptian Vulture, photograph by Mubashir Hasan

White-backed Vulture or Oriental White-backed Vulture
Gyps bengalensis
Gidh in Urdu and Hindi

The White-backed Vulture is the commonest vulture of the valley of the Indus. It is resident in the plains of all the provinces of Pakistan; is absent from central and western Balochistan as well as south-western Frontier province. It is found throughout India, Bangladesh and parts of Nepal.

Being a carrion eater with not much of a striking appearance, the White-backed Vulture is not a popular bird. However, its remarkably efficient services in keeping the town and countryside clean of carcasses, free of charge, are unmatched. Soaring most gracefully, high in the heavens for hours on end, coursing an estimated two hundred (323 kms) or more miles a day (2.1:170), hardly a dead body escapes its vigil. Once it is spotted, scores of birds will descend upon it from all directions and devour it with amazing speed.

The White-backed Vulture is a large bird with a very large wing-span. It is blackish-grey in general appearance. Its hairless neck and head are tucked away into the shoulders. Its mid-back

White-backed Vulture preparing for flight, photograph by Mubashir Hasan

is white; hence the name, White-backed Vulture. Both sexes look alike.

Body length 750-850 mm.
Wing-span 1800-2100 mm.

White-backed Vultures, photograph by Mubashir Hasan

Cinerous Vulture or Eurasian Black Vulture
Aegypius monachus
Kala Gidh in Urdu and Hindi

The Cinerous Vulture is a very large bird and is found in very small numbers in Sindh, southern areas of Bahawalpur and parts of the Salt Range and Potohar during winter. Around mid-February it migrates to the southern parts of the Afghan-Pakistan border in the Frontier province and a small adjoining area in Balochistan as well as around Quetta (2.1:177). It frequents open scrub and semi-desert country. It is rare and sparse in northern and central India and Nepal. Outside Pakistan, its breeding grounds extend from southern Europe, through the Caucasus to Tibet and eastern China (1.1:299). Now, on the list of endangered species in Europe, its demand from zoos has led to ruthless netting in Pakistan, making it scarcer and scarcer.

Cinerous Vulture, photograph by Mubashir Hasan

The Cinerous Vulture feeds exclusively on carrion. Like other vultures, it spends a considerable portion of the day soaring high in the heavens, looking for carcasses.

A big brownish-black bird larger than the White-backed, the Cinerous Vulture can have a wingspan as wide as nine and a half feet [2950 mm (2.1:176)]. Its chin, throat and back of the neck has a lining of aristocratic looking feathers. Both sexes look alike.

Body length 1000-1100 mm.
Wing-span 2500-2950 mm.

Marsh Harrier
Circus aeruginosus
Safed Sira in Urdu and Hindi

The Marsh Harrier is found throughout the Punjab and most of Sindh from mid-September to mid-March when it migrates to its breeding grounds in central and northern Asia. During the periods of migration it is also found in northern parts of the valley of the Indus. During winter it is widespread in Nepal, India and Bangladesh.

It is a bird of prey of marshes, lakes and swamps where it hunts for waterfowl as large as coots and other smaller birds. It also pirates food items from other birds of prey such as kites. Unlike most other birds of prey which build their nest on trees and ledges, the Marsh Harrier's nest is on the ground among dense reed-beds (2.1:182).

In size, as large as the Pariah Kite, the adult male Marsh Harrier has a dark brown body, a pale grey to silvery grey tail

Marsh Harrier, photograph by Mubashir Hasan

and pale rufous head, neck and breast. The female is darker than the male, its crown and throat are creamy.

Body length 480-560 mm.
Wing-span 1150-1300 mm.

Shikra, Indian Shikra or Indian Sparrow Hawk
Accipiter badius cenchroides
Shikra (Female), *Chipka* or *Cheepak* (Male) in
Urdu and Hindi

The Shikra is well known in folklore for its powerful flying, great pluck, skilful hunting technique, and dash. Hiding in a leafy tree, it pounces upon any bird of manageable size, as well as rats, squirrels, lizards, etc. It has also been observed to stampede small birds for chasing one to hunt. In olden days, it was trained to hunt partridges and quails (1:236). It is resident

Shikra, photograph by Khan Mohammad

all over the Punjab and Sindh and all of India. It visits Balochistan and parts of the Frontier province in summer.

The Shikra is ashy blue-grey above, white but densely barred with rusty brown below. It is one of the commonest of small hawks—as big as a House Crow. The female is a little larger than the male and has much browner plumage above. Both sexes look alike.

Body length 290-320 mm. (males); 300-380 mm. (females)
Wing-span 600-700 mm.

White-eyed Buzzard or Buzzard Eagle
Butastur teesa
Teesa in Urdu and Hindi

The White-eyed Buzzard is common and resident in the Punjab, Sindh and southern areas of Balochistan. It is found throughout India, Nepal, Bangladesh and Sri Lanka. It preys upon lizards,

White-eyed Buzzard, photograph by Mubashir Hasan

rats, mice, frogs, snakes and large insects and is often seen perched on electricity or telegraph lines, a boundary stone or even on the ground.

Its breeding season lasts from mid-March to May. While the male brings food to the nest, the female performs the incubation duties. The pair bond is long lasting (2:191).

The White-eyed Buzzard, slightly larger in size than the House Crow, is a greyish-brown bird of prey with a white throat and a small whitish patch on the hind neck. Both the sexes look alike.

Body length 310-360 mms.
Wing-span 800-820 mm.

Long-legged Buzzard
Buteo rufinus
Chuhamar in Urdu and Hindi

Long-legged Buzzard, photograph by Mubashir Hasan

The Long-legged Buzzard is common in Pakistan, Nepal and northern India. It breeds all across Central Asia from western Mongolia to Asia Minor and also in the Karakorams—Kaghan and Chitral valleys. It feeds on reptiles, frogs, disabled birds and dead animals; also attacks domestic chickens and pigeons. Soars high up in the sky for hours and surveys its surroundings from perches on tree tops in search of prey.

The Long-legged Buzzard is as big as the Pariah Kite; head, neck and shoulders are creamy white and the rest of the body may be pale brown, dark brown or reddish-brown. The tail is always pale chestnut and unbarred (2:193). The female is a little larger than the male. Legs are unfeathered and relatively long. Both sexes look alike.

Body length 500-650 mm.
Wing-span 1260-1430 mm.

Greater Spotted Eagle
Aquila clanga
Kaljanga in Urdu and Hindi

The Greater Spotted Eagle is found in all the riverain plains of the Punjab and Sindh, usually in the vicinity of lakes and canal headwork ponds where it is common. It is also found in Nepalese lowlands, northern India and Bangladesh. Occasional breeding activity has been reported from Sindh but most of its population is reported to migrate to Central Asia for breeding.

It feeds on rats, mice, lizards, frogs, disabled or weak birds sometimes pirated from other birds. A little larger than the Pariah Kite, the Greater Spotted Eagle is a chocolate brown or dark brown bird of prey. White specks on wing coverts are visible in sub-adult birds. Both sexes look alike.

Body length 650-720 mm.
Wing-span 1550-1820 mm.

Greater Spotted Eagle, photograph by Mubashir Hasan

Steppe Eagle
Aquila rapax
Okaab or *Ragar* in Urdu and Hindi
Parmar in Sindhi

The Steppe Eagle is commonly found all the year round, in the Punjab, Sindh and south-eastern Balochistan. It avoids mountainous areas (2:197), frequents semi-desert and dry deciduous country, chiefly plains and plateaux. It is a winter visitor to the NWFP and northern Balochistan but it is also found in Bangladesh, the Nepal Terai and the drier parts of India. Its breeding season extends from November to April. It feeds on reptiles, birds, and mammals, according to Ali and Ripley, 'mostly robbed from kites and other hawks' (1:277); also on carrion.

The Steppe Eagle which is no longer considered a species separate from the Tawny Eagle migrates to Central Asia during summer for breeding. It usually frequents larger lakes where

Tawny Eagle and Steppe Eagle, photograph by Syed Asad Ali

waders and waterfowl abound (2.1:197), and probably itself migrates northwards as they migrate.

In size, considerably bigger than the Pariah Kite, the Tawny Eagle and the Steppe Eagle vary in colour from very dark brown to pale brown to nearly whitish. Their legs are feathered up to the toes. Both sexes look alike. The females are a little larger than the males.

Body length 650-770 mm.
Wing-span (Tawny) 1690-1800 mm.
Wing-span (Steppe) 1740-2600 mm.

Imperial Eagle
Aquila heliaca
Jumiz, Bara Jumiz or *Satangal* in Urdu and Hindi

In winter, the Imperial Eagle is found in Sindh, Bahawalpur, parts of southern Balochistan and in the doab of the rivers Indus and Jhelum up to as far north as the southern areas of the Salt

Range. According to Roberts, it is resident in a small area of south-eastern Balochistan (2:199). Over a period of one hundred years, there have been reports of breeding from a few scattered points in Pakistan. However, its main breeding area is Central Asia. It is also found in north-west India and Nepal.

Imperial Eagle, photograph by Mubashir Hasan

The Imperial Eagle frequents open treeless country and is known for pirating its food by forcing other hawks and eagles to surrender their prey to it. Also feeds on carrion but will also hunt rodents and birds in the plains or along the banks of large lakes favoured by waterfowl.

A large and handsome bird of prey, as big as a vulture, the adult Imperial Eagle is blackish-brown with buff to whitish head and neck. Both sexes look alike.

Body length 720-830 mm.
Wing-span 1900-2100 cm.

OSPREYS

Family
Pandionidae

Osprey
Pandion haliaetus
Machhlimar, *Machhariya* or *Machhmanga* in Urdu and Hindi

From September to the end of April, the Osprey is found throughout the riverain areas of the valley of the Indus. It frequents large bodies of water and is absent from Balochistan and most of the NWFP. For the summer it migrates to its

Osprey, photograph by Rolf Passburg

breeding grounds in Iran and Central Asia. Ali and Ripley report evidence of breeding sites in the Himalayas, but not from any area of Pakistan (Roberts' personal communications).

The Osprey feeds exclusively on fish and is endowed with great hunting skills. It can fly or glide over water, occasionally hovering in mid-air to select and take aim at its target before plunging into the water, feet first. If successful, it emerges with the fish firmly grabbed in its talons.

The Osprey's throat, breast and belly are white, so are its crown and back of the neck; the rest of the body is dark brown. In size, it is as large as the Pariah Kite, the female being a little larger than the male. Both sexes look alike.

Body length 550-580 mm.
Wing-span 1450-1700 mm.

FALCONS

Family Falconidae

Red-headed Merlin or Red-necked Falcon
Falco chicquera
Turumti, Turumtari in Urdu and Hindi
Chatwa in Sindhi

Red-headed Merlin is a beautiful grey and white falcon with a chestnut-red head. Sexes look alike, the female is a little larger than the male.

Red-headed Merlin, photograph by Rolf Passburg

It is reported by Roberts to have drastically declined in numbers since the 1940s, due in some measure at least, to the falcon trade. It is a resident of Sindh and southern Punjab in Pakistan but is generally distributed throughout South Asia except Sri Lanka. Rare in Balochistan and southern NWFP. Breeds in early spring.

Feeds mainly on small birds which are hunted jointly by the pair.

Body length 310-360 mm.
Wing-span 600-700 mm.

PARTRIDGES

Family Phasianidae

Some of our most beautiful and popular birds are members of the family called *Phasianidae*. It comprises partridges, chakors, quails, pheasants, junglefowls and peafowls. Domesticated fowls, which are an important source of protein in human food, also belong to this family.

The *Phasianidae* are land birds with strong legs. They are expert runners and, except for quails, which are migratory, do not fly over long distances.

Most of them lay their eggs on the ground and the young are able to follow the parents soon after hatching.

Chakor or Chukar
Alectoris chukar
Chakor in Urdu and Hindi
Kakov in Kashmiri

This handsome partridge is found in the Karakorams, western and southern parts of the Frontier province and parts of Balochistan. According to Roberts, a few still survive in the Kirthar Range in Sindh, in the Salt Range around Sakesar and on the Margalla hills in the Punjab (2.1:227). Ali and Ripley quote Dr Ticehurst recording that in a good year three guns shot 700 birds in two mornings in Balochistan (1.2:19). The Chakor is a bird of the hillsides of rocky arid areas, with not much

Chakor, photograph by Khan Mohammad

vegetative cover. It is also found in the Himalayas as far east as central Nepal, and rarely at elevations as high as 5,000 metres.

Their breeding season is from April to mid-July depending upon the altitude. The normal clutch of eggs is 7 to 10 but up to 20 have been found in one nest, probably the product of two hens (1.2:20). Most of the incubating is done by the female. The Chakor is believed to be monogamous. Its chicks, reared as pets, will move freely about the house following the master and will boldly attack strangers and stray dogs. Chakor fighting is a popular sport in some parts of Pakistan and India.

Chakor is a greyish pinkish bird. Its white throat is ringed by a black band starting from its orange bill, going down the sides of the neck and completing the 'necklace' at the upper breast. There are black and buff bars on the flanks.

Body length 340-390 mm.
Wing-span 500-540 mm.

Black Partridge or Black Francolin
Francolinus francolinus
Kala Teetar in Urdu and Hindi

The Black Partridge is one of the most handsome of the birds of
the region. It is also a popular bird which can be reared as a pet.
The call of the male Black Partridge is loud and clear and
carries very far. There is something indefinably special about it.
For long, humans have attributed words of their own tongue to
the call of the Black Partridge. Urdu-speaking devout Muslims
render it as '*Subhan-teri-qudrat*' ('Praise be to the power of
Allah to create'). The Mughal Emperor Babar rendered it as
'*Shir-darem-shakrak*' ('I have milk and a little sugar') (1.2:23),
while a grocer's rendering is described as '*Lay-noon-tail-adrak*'
('Buy salt, oil and ginger').

The population of the Black Partridge has very rapidly
declined in Pakistan over the past fifty years (2.1:230). It occurs
all over the Punjab and Sindh. It feeds on shoots, leaves, grains,
insects, and larvae and not being a bird of arid, scrubless areas,
it is virtually absent from most of Balochistan and parts of the

Black Partridge, photograph by Khan Mohammad

Frontier province. It is found throughout northern India and part of Bangladesh.

The male has black lower plumage, a chestnut collar and a white patch on the cheeks, while the rest of the body is beautifully spotted, barred and scalloped with black, white and fulvous. In the female, the white patch on the cheek and chestnut collar are missing; the body is paler and browner but very beautiful.

Body length 330-360 mm.
Wing-span 500-550 mm.

Grey Partridge or Grey Francolin
Francolinus pondicerianus
Teetar, Gora Teetar, Bhura Teetar and *Ram Teetar*
in Urdu and Hindi
Achho Teetar in Sindhi
Firufti in Punjabi
Kapinja in Balochi
Tauzarai in Pushtu

Grey Partridge, photograph by Syed Asad Ali

A very common and popular bird, the Grey Partridge is much sought after by sportsmen, by gourmets, and by those who like to rear it as a pet. Properly trained it will follow its master about the house or in a jungle, where it can be made to give combative calls by professional netters to draw partridges in the area for a fight. Nets spread in the vicinity of the caller trap the challengers. Partridge fights used to be a popular sport in the olden days.

The Grey Partridge is found throughout the subcontinent—Pakistan, India, Bangladesh, Sri Lanka. It feeds on grains, berries, insects and termites.

The Grey Partridge is a greyish-brownish bird, heavily barred and has spots and blotches of several shades and colours. The tail is pale chestnut. Both sexes look alike.

Body length 330-350 mm.
Wing-span 480-520 mm.

WATERHENS, MOORHENS
─── AND COOTS ───

Family
Rallidae

Except for the coot which regularly frequents open water like a duck, the members of the *Rallidae* family are birds of wetlands endowed with thick vegetative cover. Frequenting marshes, banks of lakes and backwaters of canal headworks, these shy, almost secretive ground-dwelling birds are rarely seen in the open. Some of them walk with the jerky gait of a domesticated hen. They like to probe in mud or shallow muddy water and their bills are well-adapted for this, their diet comprising mostly of aquatic insects and their larvae, worms, molluscs, etc. They will also eat seeds of sedges, berries and green shoots or succulent roots to availability.

White-breasted Waterhen
Amaurornis phoenicurus
Dawak, Kinati or *Dahak* in Urdu and Hindi
Kuraki in Sindhi

Whistler describes the White-breasted Waterhen as a very noisy bird remarkable for its sharp metallic sounding calls 'much like the noise of pounding with pestle and mortar' often kept up all night long (4:437). It is commonly found skulking in the vicinity of lakes, ponds, forest plantations and gardens, always within easy access of a dash into cover of shrubbery and vegetation. Searching for insects, worms, seeds and shoots of vegetation, it

White-breasted Waterhen, photograph by Mubashir Hasan

keeps jerking its tail up as it walks. It loathes flying but is known for its agility to clamber reeds and thorny bushes.

Larger than the myna and smaller than the pigeon, the White-breasted Waterhen is a slaty grey bird with a short tail and white breast and cheeks. Both sexes look alike. It is common in Sindh and in some parts of the Punjab but is absent from the NWFP and Balochistan. It is found throughout northern India, Sri Lanka and parts of Nepal.

Body length 320 mm.
Wing-span 450-540 mm.

Moorhen or Watcrhen
Gallinula chloropus
Jal Murghi or *Pani Murghi* in Urdu and Hindi

The Moorhen is a slaty grey and brown bird smaller than a pigeon but larger than a myna. It has a bright red forehead and a white patch under the tail and it shows a streaked white line along the flanks. Both sexes look alike.

Moorhen, photograph by Mubashir Hasan

The Moorhen may be seen all the year round throughout the valley of the Indus along the banks of lakes, ponds, backwaters of canal headworks and other bodies of water which have vegetative cover. In the rest of the subcontinent it is found throughout India, Bangladesh, Nepal and Sri Lanka. Ali and Ripley describe it as resident with a part of its population migrating to Central Asia for the summer (1.2:175). The Moorhen feeds on land as well as in water—on seeds, vegetable matter, frogs, small fish, insects and molluscs. It is a shy bird and does not wander far from the cover of reeds and shrubbery.

Body length 300-330 mm.
Wing-span 490-530 mm.

Purple Gallinule, Purple Moorhen, Purple Swamphen or Purple Coot
Porphyrio porphyrio
Kalim, Kharim or *Jalmurgha* in Urdu and Hindi
Wan Tech in Kashmiri

Of the size of a domestic hen, the Purple Gallinule is a brightly coloured bird commonly found, all the year round, on all large lakes and backwaters of canal headworks in Pakistan; also found throughout India.

The Purple Gallinule mainly feeds on vegetable matter but also eats insects and molluscs. It wades over floating vegetation or feeds in the vicinity of reeds along edges of lakes constantly raising and lowering its tail.

The naked bill and forehead are bright red and legs a duller red. The body is glistening purple and throat bluish. In strong sunlight, the wings and shoulders appear distinctively green. Both sexes look alike and they breed during summer.

Body length 450-500 mm.
Wing-span 900-1000 mm.

Purple Gallinule, photograph by Mubashir Hasan

Eurasian Coot
Fulica atra
Dasari, Dasarni, Aari, Khuskul or *Coot* in Urdu and Hindi
Aari in Sindhi
Kolur or *Kolru* in Kashmiri

Slightly larger than a pigeon, the Coot, for all practical considerations, is a waterfowl like any other duck. It prefers open waters and dives for its food. It is found in very large numbers in all the large bodies of water in Pakistan, India, Nepal, Bangladesh and recently Sri Lanka. During March/April, it leaves for its breeding grounds in Central Asia but also breeds in the Himalayas, Karakorams and in the lakes in Balochistan and Sindh (2.1:267).

The Coot feeds on vegetable matter, insects, worms and molluscs. It is an all black bird with an off-white bill and crimson iris. Both sexes look alike. Unlike a duck, it can take off from water only with effort, but once airborne it is capable of sustained long distance flight for migration over high mountains. A bird ringed in India was recovered in Uzbekistan 2,240 km away. Another ringed in Kazakhstan was recovered in Kashmir 1,600 km away (1.2:181).

Body length 360-380 mm.
Wing-span 700-800 mm.

Eurasian Coot, photograph by Mubashir Hasan

CRANES

Family Gruidae

Common Crane
Grus grus
Koonj in Urdu, Hindi and Sindhi

A large grey bird with long neck and legs, the Common Crane may stand as high as 1400 mm (4.5 feet). It is a popular bird known for its migratory flight patterns, high-pitched calls and wariness of humans. The Common Crane is found in Pakistan mainly on its passage to and from India. It leaves for its breeding

Common Crane, photograph by Syed Asad Ali

grounds in Siberia and Northern Europe in March/April and returns in September/October. In Pakistan it is mercilessly hunted and persecuted. Surveys of crane hunting in 1982 and 1983 revealed that up to 750 Common Cranes are captured alive or killed in the Kurram valley alone. According to one estimate there were at least 5,170 captive cranes in the NWFP in early 1983 (2.1:270). Little wonder that cranes are very wary of humans. To select a safe area for landing, they circle over it for a long time and land only in the middle of a large exposed plain generally near bodies of water which permits them to see an approaching human from a great distance. Cranes are terrestrial in habit and like wild and remote country for breeding. They pair for life (11:161).

Cranes migrate at extremely high altitudes and always in very large numbers. The characteristic call of cranes is loud and trumpet like. The calls of the alarmed cranes make a tremendous din which has been likened to the distant roar of the sea (1.2:138). Cranes rest during the day and fly out to feed in the mornings and evenings mainly on cereal crops, tubers, or shoots of grass. They also eat insects, reptiles and small birds (11:161).

The head and neck of the Common Crane is blackish except for a broad white band going down along the sides of the neck. The rest of the body is grey. The innermost secondaries are developed into filamentous plumes which cover the tail and the wingtips.

Wing-span 2200-2450 mm.
Height 1100-1200 mm.

BUSTARDS

Family
Otididae

All the five members of the *Otididae* family found in Pakistan are in danger of extinction due to indiscriminate hunting and netting. They are long necked and long-legged birds of desert or semidesert habitat and vary in size from that of a village hen to a vulture. They are expert at running and hiding but also have strong wings suitable for long distance migration to their breeding grounds.

Likh or Lesser Florican
Sypheotides indica
Likh, Chhota Charaz, Barsati or *Kala Charaz* in Urdu and Hindi
Khar Mur or *Tan Mur* in Sindhi

The Likh is one of the smaller bustards. It is a bird of the subcontinent inhabiting open grasslands and hence migratory according to the incidence of monsoon rains. It feeds on shoots of crops and grasses, seeds and berries and on insects of all kinds. It visits in Pakistan, during the rainy season, the southern-most parts of Sindh. Roberts has reported rare sightings from Kasur and the Sukh-Beas area (2.1:281). In India it is found in the south and spreads northwards during the rainy season, hence its Urdu name *Barsati*. In breeding season, the male displays to attract a mate by jumping in the air to a height of one to one and a half metres, the jump lasting only two to three seconds in duration.

Likh, photograph by Khan Mohammad

The Likh has sandy buff upper parts mottled with blackish arrowhead marks and irregular fine black lines. Its throat and chin are white. In winter the sexes look alike, the female being slightly bigger than the male. In breeding season the head, neck and breast of the male turn black and it develops a long crest of spatulate black plumes at the back of the neck.

Body length 400-480 mm.

JACANAS

Family
Jacanidae

Pheasant-tailed Jacana
Hydrophasianus chirurgus
Piho or *Pihya* in Urdu and Hindi
Gund Kav in Kashmiri

The Pheasant-tailed Jacana is a crow-sized bird with a long curving tail in the summer season. It frequents large bodies of

Pheasant-tailed Jacana, photograph by Khan Mohammad

water as well as inundated roadside burrow-pits. It walks from leaf to leaf searching for insects and larvae, molluscs and amphibia. As one leaf begins to sink due to its weight, it lightly steps on to another. In Pakistan it is found in Sindh all the year round but is only a summer visitor to Punjab. It is found throughout India, Bangladesh and Sri Lanka.

Jacanas are polyandrous. The female leaving the eggs with one male for incubation, proceeds to help another in building his nest and perform his duties of incubation and rearing the chicks. Several clutches of eggs may thus be laid and several families produced by a single female.

During the monsoon, the Jacanas wear breeding plumage which may last from early May to early October (2.1:282). They develop pheasant-tails, black breasts and a bluish bill. In winter the pheasant tail disappears, the breast becomes white and the bill grey-green.

Body length (females in breeding plumage) 500-540 mm.
(males in breeding plumage) 440-470 mm.

SNIPES

Family Rostratulidae

Birds of this family are found throughout South East Asia, South Asia and Central Africa. The males are slightly smaller than the females, the male doing all the incubating and caring for the young whereas the females are polyandrous. Adapted to swampy burrow pits and reed-fringed ponds or stagnant drainage channels choked with reeds, they are resident throughout the Indus basin. They tend to occur in small colonies usually around rice fields. They feed on a variety of aquatic insects and their larvae, as well as weeds of aquatic plants. They sometimes feed in shallow water by scything their bills to and fro.

Painted Snipe
Rostratula benghalensis
Rajchaha in Urdu and Hindi

A little larger than the myna, the Painted Snipe is not a true snipe except in looks. It flies slowly with dangling legs whereas the flight of a snipe is fast and dashing. Besides, it is a multi-coloured bird, the female more so than the male. It is found throughout the year in irrigated areas of the Punjab and Sindh wherever there is thick vegetative cover over a perennial patch of water such as cover swamps, jheels, burrow-pits and stagnant drainage pools. In similar habitat it is found throughout India, Bangladesh and Sri Lanka.

Painted Snipe, photograph by Syed Asad Ali

Throughout the year, the female's chin, throat and upper breast are rich maroon chestnut. Upper parts are chiefly metallic olive or bronze with buff and blackish streaks and markings. It also wears a whitish 'spectacle' and is larger than the male. The male lacks chestnut and black on the neck and breast. It is the male who incubates the eggs and brings up the family. After laying a clutch of eggs the polyandrous female moves on to help lay another and yet another. Both sexes look alike.

It feeds on insects, worms, crustaceans, molluscs, seeds and vegetable matter.

Body length 230-260 mm.
Wing-span 500-550 mm.

― OYSTER-CATCHERS ―

Family
Haematopodidae

Oyster-catcher
Haematopus ostralegus
Darya Gajpaon in Urdu and Hindi
Dobah in Sindhi

The Oyster-catcher is found on the seacoast of Karachi and Makran from September to May. Some individuals stay on during summer, while most migrate by mid-May to their

Oyster-catcher, photograph by Khan Mohammad

breeding grounds, presumably along the coasts of the Black and Caspian seas, although no migration data is available (2.1:287). In India they are commoner on the west coast than the east.

The Oyster-catcher is a partridge-sized, black-and-white bird with coral pink legs and an orange bill. Its wedge tipped bill is specially adapted for prying open hard-shelled marine life. It feeds on oysters and other bivalve molluscs. Both sexes look alike.

Body length 400-450 mm.
Wing-span 800-860 mm.

STILTS, AVOCETS

Family Recurvirostridae

Recurvirostridae is a family of birds with long thin bills and very long torsos and tibia giving its members extra long legs, and on that account two of its most common specie are called stilts. Of the two species found in the Indus Valley, *himanotopus* has a straight bill. They wade in shallow water and probe for food with bills fully submerged often upto their eyes, or in deeper water upto their breasts. Their food comprises mainly of animal matter including larvae, pupae, aquatic insects, water-fleas, etc. and to a lesser extent seeds of rushes and sedges.

Black-winged Stilt
Himantopus himantopus
Gaz Paon in Urdu and Hindi
Gusling in Sindhi

The epithets, Stilt, in English, and *Gaz Paon*, meaning 'yard-long feet', in Urdu and Hindi refer to the enormously long legs of the Black-winged Stilt. It is an elegant pigeon-sized marsh bird with a white body and black wings.

On the edges of bodies of shallow water, of large lakes or of small roadside burrow-pits, the Black-winged Stilt is a very common bird. It is a summer visitor to northern Punjab for breeding (2.1:290). In the rest of the Punjab and all of Sindh, it is found throughout the year. It is found throughout the NWFP and Balochistan where there is suitable water. It is a resident of

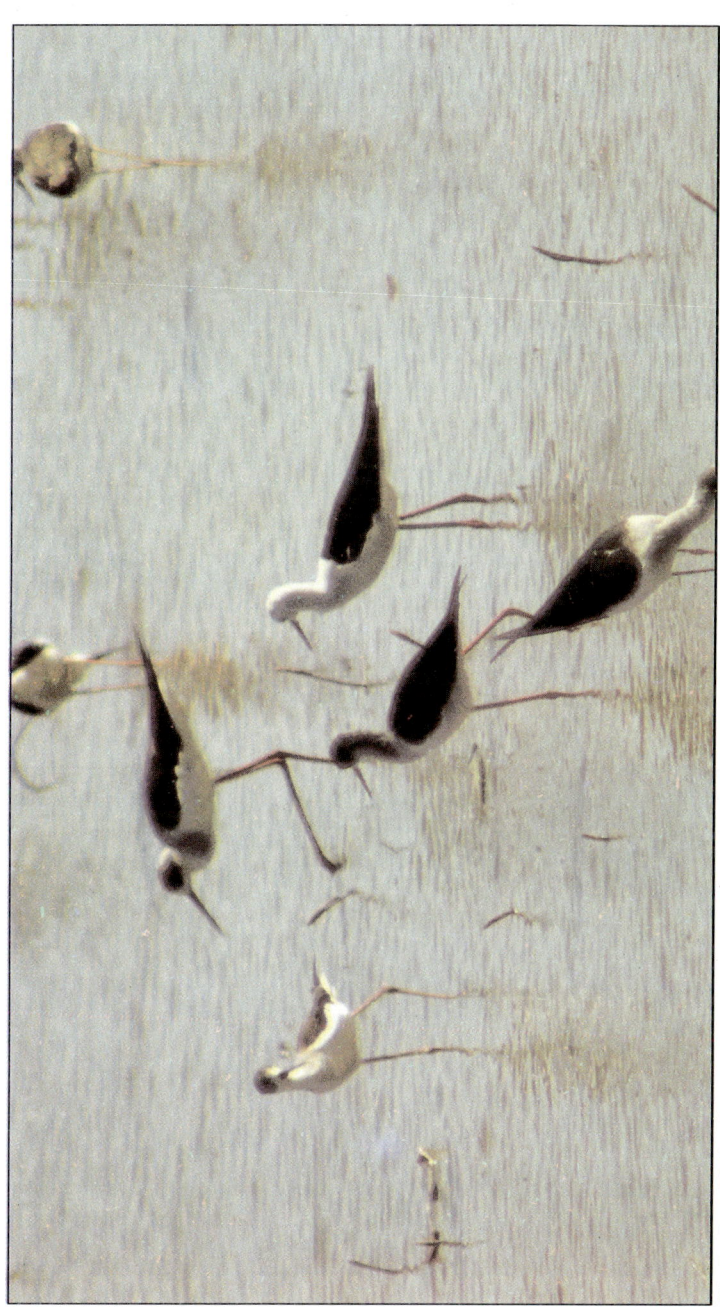

Black-winged Stilt, photograph by Mubashir Hasan

the Khusdil Khan lake in Balochistan (Roberts p.c.). It is found throughout India, Nepal and Bangladesh. It feeds on insects, worms, crustaceans, molluscs, seeds and vegetable matter.

Body length 350-400 mm.
Wing-span 670-830 mm.

Avocet or Pied Avocet
Recurvirostra avosetta
Kusya Chaha in Urdu and Hindi

The Avocet is found in small numbers throughout the year. However, for reasons not clear, there is a noticeable increase in the numbers from spring to early summer (2.1:291). By the end of May most leave for their breeding grounds probably in Iran, Afghanistan and Central Asia, but some stay on to breed in the Rann of Kutch and Balochistan. It is common in north-west Nepal and India but not so in the rest of the Union.

Avocet, photograph by Syed Asad Ali

Avocet is a beautiful black-and-white pigeon-sized bird with longish legs frequenting salt or fresh water marshes, creeks or edges of drying out ponds. Like bold strokes from the brush of a superb artist, its snow white body carries several bold black stripes on the wings as well as back. There is one stretching from the base of the bill over the crown and the back of the neck. It is unique in having an up-curved bill which distinguishes it from any other bird (1.2:332). Both sexes look alike. It feeds on insects, tiny crustaceans, and molluscs.

Body length 420-450 mm.
Wing-span 770-800 mm.

STONE CURLEWS
── OR THICK-KNEES ──

Family
Burhinidae

Burhinidae is the family of curlews which are plover-like birds found throughout northern India, Aghanistan and the Middle East. They are cursorial birds, with thick-kneed legs and huge 'goggle' eyes. Being largely nocturnal in feeding activity, their large eyes probably indicate a high degree of visual accuracy in darkness. They usually associate in pairs or small groups during the winter. They subsist on a variety of insects like grasshoppers, worms, molluscs, amphibia, etc.

Great Stone Plover
Esacus recurvirostris
Qurbanak or *Bada Karwanak* in Urdu and Hindi

The Great Stone Plover is found all the year round on the seacoast in Lasbela and along the mouths of the Indus and inland at a few places along the rivers Chenab, Jhelum and the Indus. It is also found in India, Bangladesh, Nepal, Sri Lanka, Burma and Vietnam.

A long-legged bird, little bigger than a kite, it is greyish-sandy above and white below. Both sexes look alike. It prefers stony drying out shores or shingly islands. It subsists on animal food such as crabs and other crustaceans, insects, molluscs, frogs, etc. and is equipped with a strong heavy-looking bill

Great Stone Plover, photograph by T. J. Roberts

suited for upturning stones and procuring items of its diet from underneath.

Body length 480-520 mm.
Wing-span 900-1000 mm.

COURSERS,
────── PRATINCOLES ──────

Family
Glareolidae

Cream-coloured Courser
Cursorius cursor
Askalo in Bruhui

The Courser is a sandy buff lapwing-like, patridge-sized bird.
There is a rufous and ashy-grey crown with a broad black-and-

Cream-coloured Courser, photograph by Rolf Passburg

white stripe from behind the eye to the nape. Sexes look alike. According to Roberts, if not molested they will become very tame and not fly away on approach by humans.

The Courser may be found in Sindh and Balochistan where it is resident. It is a winter visitor to arid areas of the Punjab. It feeds chiefly on caterpillars, beetles, ants and other insects.

Body length 190-220 mm.
Wing-span 510-570 mm.

Small Sindhi Pratincole
Glareola lactea
Utteran in Sindhi

The Small Sindhi Pratincole arrives in Pakistan from East Africa. It breeds here and in many parts of India. It has been sighted from February to September (2.1:301). Roberts reports the riverain areas of the Indus from the sea to the Taunsa Barrage and of the rivers Sutlej and Ravi as its breeding grounds. It feeds on insects, beetles, bugs, termites and winged white ants.

Small Sindhi Pratincole, photograph by Mubashir Hasan

The Pratincole is a myna-sized olive-brown bird with a creamy, yellowish throat bordered by a black band. The underwings are chestnut, visible when the bird raises its wings when at rest (5:142). Both sexes look alike.

Body length 230-250 mm.
Wing-span 600-650 mm.

— PLOVERS, LAPWINGS —

Family
Charadriidae

Members of the family *Charadriidae* are short-billed birds
varying in size from bulbul to pigeon. They are terrestrial waders
of open habitat. They stand upright with head held high. Males
are slightly larger than females. Both the sexes look alike or
nearly so (16:113).

The wing-tips of lapwings are always black and usually have
a broad white bar. These birds are generally never far away
from water. They prefer to run rather than walk. They go a
short distance and then stop, hold their head up, as if listening,
peck a little before undertaking the next run. According to
Roberts the members of this family found in Pakistan are thirteen
in number (2.1: 36). Seven are included in this volume.

Little Ringed Plover
Charadrius dubius
Merwa or *Zirrea* in Urdu and Hindi
Kola Katij in Kashmiri

A dainty brown and white bulbul-sized bird with a sharply
defined black collar around its neck, the Little Ringed Plover
frequents wet banks of rivers, lakes and ponds all over Sindh
and the Punjab and parts of the NWFP and Balochistan. It is
found throughout the subcontinent and further east up to
Vietnam. During breeding season it wears a conspicuous yellow
orbital ring. Both sexes look alike. It runs about in short spurts

Little Ringed Plover, photograph by Syed Asad Ali

suddenly stopping to peck at insects, worms, tiny crabs, etc. It is also known for rapidly tapping uneven ground to stampede tiny insects and crustaceans in order to catch them (1.2:233).

Body length 140-150 mm.
Wing-span 420-480 mm.

Kentish Plover or Snowy Plover
Charadrius alexandrinus

A dainty light brown and white bulbul-sized bird, the Kentish Plover frequents not only wet banks of rivers, lakes and ponds but also sand-dunes (2.1:308). It breeds along the coast of Pakistan and at a few places in the interior of Sindh, Punjab, Balochistan and in northern India. In non-breeding season it spreads all over.

Kentish Plover, photograph by Mubashir Hasan

This plover is similar to the Little Ringed Plover but is without any collar around its neck. Both sexes look alike. It feeds on insects, worms, tiny crabs, etc.

Body length 150-170 mm.
Wing-span 420-450 mm.

Lesser Sand Plover or Mongolian Plover
Charadrius mongolus

The Lesser Sand Plover is a wader and looks closely similar to the Kentish Plover (C. Aledandrinus, Roberts, 1991) except for its slightly larger size and lack of the narrow white collar on the back of the neck. It may be seen in winters along the Sindh and Balochistan coast in very large numbers often mixed with other waders, dunlins, stints and Large Sand Plovers. It breeds in high mountainous plateaus of the Karakorums and Himalayas.

Its food consists of baby crabs, sandhoppers and marine worms.

Lesser Sand Plover, photograph by Rolf Passburg

Both sexes look alike.

Body length 190-200 mm.
Wing-span 450-480 mm.

Large Sand Plover or Greater Sand Plover
Charadrius leschenaultii

The Large Sand Plover is found from August to May on the
seacoast of Pakistan. It is also found on the seaboards of India
and Bangladesh. Its breeding grounds lie in Central Asia
although no ringing data is available. It is a myna-sized bird of
not a very striking appearance in its winter plumage. Ashy-
brown above, white below, it frequents sandy beaches pecking
at baby crabs, beetles and insects. Both sexes look alike.

Body length 220-250 mm.
Wing-span 128-140 mm.

Large Sand Plover, photograph by Mubashir Hasan

Yellow-wattled Lapwing
Hoplopterus malabaricus
Zirdi in Urdu and Hindi
Jithiri in Sindhi

The Yellow-wattled Lapwing is so called for its broad triangular shaped flaps of yellow fleshy skin, as if hanging from the base of the black tipped bill, and also extending in front of the forehead. This lapwing, hailing from peninsular India, visits the southern-most part of Pakistan around Karachi and Lasbela during summer for breeding.

It is a slightly smaller and slimmer version of the Red-wattled Lapwing (2.1:315) but unlike the latter it prefers dry fallow land and is not dependent upon the proximity of water. Both sexes look alike. It feeds on insects, grasshoppers and beetles.

Body length 240-280 mm.
Wing-span 650-690 mm.

Yellow-wattled Lapwing, photograph by Syed Asad Ali

Red-wattled Lapwing or Red-wattled Plover
Hoplopterus indicus
Tateeri, Titi or *Titori* in Urdu and Hindi
Tateehar in Sindhi
Hatatut in Kashmiri

Its food supply of beetles, ants, caterpillars, grasshoppers, snakes, worms, molluscs, larvae and vegetable matter are commonly available in the plains of the Indus, so the Red-wattled Lapwing is also almost everywhere—in the countryside as well as in the open spaces of towns, all the year round. It feeds in shallow water as well as in open fields. Acknowledged throughout for its sharp and strong teet-teet-tit-toowhit calls and bold presence, it is equally at home in crowded city parks as in deserted seepage zones. It is a pigeon-sized bird with long legs and is found throughout the subcontinent.

In the jungle it keeps a watch over the environment by promptly announcing the presence of a sportsman or a photographer trying to stalk a bird or an animal. It can be a noisy bird. In summer it can call all night long. Before the

A partial albino Lapwing

Red-wattled Lapwing, photograph by Mubashir Hasan

arrival of monsoons the farmer attributes these calls as a request to God for rain. For reasons not clear, the Red-wattled Lapwing has earned the envy of humans who, in the Punjab, have concocted a story that this lapwing sleeps with its legs pointing upwards. The implied ridicule is then added 'to stop the sky from falling on the Earth'.

The Red-wattled Lapwing breeds in summer. One day as this writer was walking along the bank of the Jhelum River at Trimmu, a lapwing landed just ahead of him. He learnt as he followed the bird that he was being taken from the path which its little chicks were crossing to scurry into a bush.

The Red-wattled Lapwing has a red flesh wattle in front of each eye. It is an olive-brown bird; a broad pure white band separates the brown from its black head, throat and upper breast. The belly is white. Both sexes look alike. This author, however, once photographed a member of the species which was a partial albino and had a white upper body instead of brown. The bird seemed to move about among other lapwings in a normal way.

Body length 320-350 mm.
Wing-span 800-810 mm.

White-tailed Lapwing
Chettusia leucura

The White-tailed Lapwing is a myna-sized bird with long legs. It frequents marshy areas and lake shores and feeds in shallow water as well as on the banks on molluscs, worms and water insects. From September to March it is found in the riverain plains of the Punjab and Sindh. It is also found in north-western India and in smaller numbers in Bangladesh and the rest of northern India. It passes through Balochistan on its way to and from its breeding grounds from Afghanistan to Iraq and parts of Central Asia. On one occasion a nest and eggs were also found in the Zangi Nawar lake in Balochistan (2.1:320 and 1.2:206).

White-tailed Lapwing, photograph by Mubashir Hasan

The White-tailed Lapwing is a light brown plover; belly and tail are white and legs are yellow. Both sexes look alike.

Body length 260-290 mm.
Wing-span 670-700 mm.

Green Plover, Peewit or Lapwing
Vanellus vanellus

The name Peewit is derived from its call, pee-wit, pee-wit. From October to March it frequents grassy shores and edges of lakes and other bodies of water in most of the Punjab, parts of the NWFP and a few areas of Sindh. In India it is found in the north-west and in smaller numbers in the north. It breeds in Europe and North Asia and feeds on molluscs, worms and insects.

Green Plover, photograph by Mubashir Hasan

The Green Plover is a partridge-sized bird and is conspicuous for its long upstanding crest. The body has a generally black-and-white appearance. Both sexes look alike.

Body length 310 mm.
Wing-span 220-236 mm.

— SANDPIPERS, STINTS —

Family
Scolopacidae

Sanderling
Calidris alba

Wherever there are sandy beaches in the world there is the Sanderling. It is found from September to May (2.1:325) in Sindh and Balochistan, all along the seacoast of Pakistan as well as of India, Bangladesh and Sri Lanka. For breeding it migrates to the Arctic region of the Old as well as the New World. The distances covered by it to and from its breeding grounds are phenomenal, such as, from the southern-most tip of Africa to the North Sea or Greenland.

Sanderlings and Dunlins, photograph by Mubashir Hasan

It feeds on beaches. Like a computerized toy, avoiding advancing waves, it runs and stops and runs again at the water's edge pecking and probing for food which consists of crustaceans, molluscs and other tiny animals washed up on the seashore.

In size, as large as a bulbul, the Sanderling has a white face, pale grey back and a blackish shoulder patch. Under parts are white. As it dons breeding plumage by mid-May, the upper parts turn light chestnut mottled with black and head, neck and breast light chestnut with darker streaking. Both sexes look alike.

Body length 200-210 mm.
Wing-span 400-450 mm.

Little Stint
Calidris minuta
Chhota Panlawwa in Urdu and Hindi

Of the size of a sparrow, the Little Stint is the smallest of the birds found on the edges of waters, large or small, from seacoast to roadside burrow-pits. It can be seen in Pakistan from July to

Little Stint, photograph by Mubashir Hasan

April. It migrates to the arctic region for breeding. It is widespread in Sindh and the Punjab and in Balochistan but chiefly along the coast, often in the company of Dunlin and Curlew Sandpipers. It is absent from the hilly areas. In similar habitat it is widespread in India, Bangladesh and Sri Lanka.

It is a greyish bird with white under parts, with darker mottling on the head and black legs in contrast to the olive-green legs of the very similar Temminck's Stint. Both sexes look alike. Its food consists of crustaceans, molluscs, larvae, and other tiny animals like flies.

Body length 120-140 mm.
Wing-span 340-370 mm.

Temminck's Stint
Calidris temminckii
Chhota Panlawwa in Urdu and Hindi

Wherever there is a pool of water in the Punjab, there is the sparrow-sized Temminck's Stint. It feeds on crustaceans,

Temminck's Stint, photograph by Syed Asad Ali

molluscs, larvae, and other tiny insects like flies. It is less common in Sindh (2.1:327).

A tiny water bird, with dark grey upper parts and grey and brownish upper breast, the Temminck's Stint is like a miniature Common Sandpiper. Both sexes look alike. By April-May, it leaves for the arctic regions of Scandinavia and Russia and begins its return migration in August.

Body length 130-150 mm.
Wing-span 340-370 mm.

Ruff
Philomachus pugnax
Gehwala Bagbad in Urdu and Hindi

Ruff and Reeve, as the male and female of this bird are respectively called, pass through Pakistan in spring and autumn on their migratory passage enroute chiefly to the seacoasts of India and Sri Lanka. A few males can be seen in partial breeding plumage in Pakistan in March and early April but the bulk of

Ruff, photograph by Mubashir Hasan

the birds sighted in Pakistan throughout the plains of the Punjab and Sindh are in their winter plumage of a greyish-brown wader with bold blackish scaly-patterned upper parts and the bill of a sandpiper. Birds with red legs can be confused with the Redshank.

Of the size of a pigeon, the Ruff is a somewhat unique bird in the sense that it acquires during breeding season an amazingly variable plumage, almost any combination of black, white, rufous brown and buff with bars and streaks. The bill could be red, yellow, brown or blackish; legs could be green, yellow, orange or flesh coloured. It also develops conspicuous ear tufts on the back of the neck and a frill of long erectile feathers highly variable in colour around the neck. The female, called Reeve is markedly smaller and thinner necked.

It feeds on worms, insects, molluscs and vegetable matter—grass, weed seeds, rice and like other waders it is found on the edges of waters large or small.

Body length 260-300 mm.(males), 200-240 mm.(females)
Wing-span 540-580 mm.(males), 480-520 mm.(females)

Common Snipe or Fantail Snipe
Gallinago gallinago
Chaha in Urdu and Hindi
Lik Pakhi in Sindhi

Known for its lightning zigzag flight, the Common Snipe was a much hunted bird for British sportsmen before Independence. Research into old records by Roberts (2.1:337) has revealed that a bag of 200 snipes between 5 or 6 guns was not unusual and that the maximum hunting bag of 300 snipes shot in 6 hours by 3 guns was recorded in Lower Sindh in 1923.

The Common Snipe frequents all water bodies with short vegetation and is found throughout the plains of the Punjab and Sindh. It is also common throughout India, Nepal, Bangladesh

Common Snipe, photograph by Syed Asad Ali

and Sri Lanka. By the end of April it leaves for its breeding grounds from central to far north of Asia. It also breeds in Europe and the Himalayas. There is no record of breeding in Pakistan though it breeds in the valley of Kashmir. The return migration begins in August. Worms, larvae and tiny molluscs are the main items of its diet.

With a straight 60 mm long bill, the Common Snipe is a little larger than a myna. It is a brownish bird heavily streaked with black, rufous, and buff; whitish below. It is perfectly camouflaged in its marshy haunts and is very difficult to spot as it has a habit of freezing when approached. Both sexes look alike.

Body length 250-270 mm.
Wing-span 440-470 mm.

Black-tailed Godwit
Limosa limosa
Gudera or *Gairiya* in Urdu and Hindi
Susling in Sindhi

The Black-tailed Godwit is a large, crow-sized, darkish sandy-brown wader, with long legs which frequents, from September to April, mostly large fresh water bodies of Sindh and is only rarely sighted in the Punjab (2.1:344). It is also found in northern India but is rare in the south (1.2:249). It breeds in northern and central Europe and west Asia up to the Aral Sea; also in eastern China and Korea.

The Black-tailed Godwit has a long straight bill and a black-tipped tail. The long bill is used to probe into soft mud for small crabs and molluscs. Both sexes look alike, the female being larger than the male.

Body length 400-440 mm.
Wing-span 700-820 mm.

Black-tailed Godwit, photograph by Mubashir Hasan

Bar-tailed Godwit, photograph by Syed Asad Ali

Bar-tailed Godwit
Limosa lapponica
Gudera or *Gairiya* in Urdu and Hindi
Susling in Sindhi

The Bar-tailed Godwit is not easily distinguishable from its cousin the Black-tailed Godwit. Its tail is cross-barred grey and white against the black of the former and its body has a more streaked appearance. Both sexes look alike.

The Bar-tailed Godwit is found in India and Pakistan mostly along the seacoast from August to April. It breeds in the far north of Asia and Europe and winters in Pakistan along the seacoast.

Body length 370-390 mm.
Wing-span 700-800 mm.

Curlew or Eurasian Curlew
Numenius arquata
Goar, Goungh, Bara Gulinda in Urdu and Hindi
Borindo in Sindhi

The Curlew may be found in Pakistan, from September to April, on all large lakes and canal headworks in Pakistan but is abundant only on the seacoast. In similar habitat it is found all over India, south of the Himalayas and further east up to Vietnam. It breeds throughout Siberia and northern Europe.

The Curlew feeds on molluscs, crustaceans, mudskippers, insects and occasionally berries (1.2:248). It blends in well with the open mudflats where it feeds and is difficult to sight. It is a wary bird which takes to flight not allowing close approach by humans.

The Curlew is the largest among the waders. Darkish sandy brown, scalloped with fulvous above and streaked on the whitish under parts, it is as big as a kite and is equipped with a 130-170 mm long down-curving bill. Both sexes look alike, the female being larger than the male. The long bill is used to probe into

Curlew, photograph by Mubashir Hasan

soft mud for small crabs and molluscs. They feed both during the daytime and at night.

Body length 500-600 mm. inclusive of the bill
Wing-span 800-1000 mm.

Redshank
Tringa totanus
Chhota Batan in Urdu and Hindi

The Redshank is slightly larger than the myna. It is a greyish-brown wader spotted with dark brown marks. It frequents seacoasts, banks of rivers, lakes and tidal creeks and is found in all the riverain areas and big lakes of Pakistan and in similar habitat in India, Bangladesh and Sri Lanka. By the end of April, it leaves for its breeding grounds in Central Asia and begins to return in September.

The Redshank has orange-red legs and an orange-red and black bill. Its upper plumage is grey-brown, spotted with white

Redshank, photograph by Syed Asad Ali

and buff; the lower plumage is white. Both sexes look alike. It feeds on insects, worms, molluscs, crustaceans and larvae.

Body length 270-290 mm.
Wing-span 590-660 mm.

Marsh Sandpiper
Tringa stagnatilis

The Marsh Sandpiper is a myna-sized greyish-brown wader spotted with dark brown marks. It frequents banks of rivers, lakes and tidal creeks and is found in all the riverain areas and big lakes of Pakistan and in similar habitat in India, Bangladesh and Sri Lanka. By the end of April, it leaves for its breeding grounds in Central Asia and begins to return in September. A bird ringed in India was recovered 4,800 km away in the former USSR (1.2:262).

The lower plumage of the Marsh Sandpiper is white, the bill is straight and thin and the legs are green. The body is streaked

Marsh Sandpiper, photograph by Mubashir Hasan

with darkish spots. It feeds on insects, worms, molluscs, crustaceans and larvae.

Body length 220-240 mm.
Wing-span 550-590 mm.

Greenshank
Tringa nebularia
Tantana or *Timtima* in Urdu and Hindi

The Greenshank, a little bigger than the partridge, is the largest of the sandpipers of our region. It frequents lakes, swamps, seepage zones, river banks and is found in abundance throughout the plains of the Punjab and Sindh in Pakistan. It is also found throughout India, Bangladesh and Sri Lanka. During April, it begins to leave for its breeding grounds in Central Asia and begins to return in August. But it can be sighted in all months of the year (2.1:354).

The Greenshank is a dark greyish-brown bird with a whitish head and breast. The body is streaked with darkish spots. The

Greenshank, photograph by Mubashir Hasan

legs are green and the bill is slightly upturned. It feeds on insects, worms, molluscs, crustaceans and larvae.

Body length 300-330 mm.
Wing-span 680-700 mm.

Wood Sandpiper
Tringa glareola
Chupka, Chobaha or *Titvari* in Urdu and Hindi

The Wood Sandpiper is a myna-sized greyish-brown wader spotted and marked with white. It frequents banks of rivers, lakes and tidal creeks and is found in all the riverain areas and big lakes of Pakistan and in similar habitat in India, Bangladesh and Sri Lanka. By April-May, it leaves for its breeding grounds in the far north of Asia and begins to return in August. One individual ringed near Calcutta in 1967 was recovered forty-eight days later, 6,200 km (over 3,800 miles) away, in the Magadan region of Siberia (1.2:266). An isolated population nests in the Pamirs (3,750 metres altitude). It feeds on insects,

Wood Sandpiper, photograph by Khan Mohammad

worms, crustaceans, molluscs and can be commonly seen throughout the irrigated areas of the Punjab and Sindh wherever there is an accumulation of water.

The upper plumage of the Wood Sandpiper is greyish-brown. It is spotted above and pale dusky and white below. Both sexes look alike and both share the incubation duties, the female often leaving the male to care for the young (2.1:357).

Body length 190-210 mm.
Wing-span 560-570 mm.

Common Sandpiper
Tringa hypoleucos

The Common Sandpiper is a bulbul-sized wader with a uniformly olive-grey upper plumage and white belly. It frequents lakes, rivers, ponds, roadside borrow-pits and can be seen over most of the Punjab and Sindh and parts of the NWFP and Balochistan. It feeds on crustaceans, molluscs and insects.

Common Sandpiper, photograph by Mubashir Hasan

The Common Sandpiper breeds in Europe, Asia, north of Iran and suitable rivers throughout the northern areas of Pakistan (2.1:359). It leaves for its breeding grounds in May to return in August. Both sexes look alike.

Body length 190-210 mm.
Wing-span 380-410 mm.

GULLS

Family
Laridae

Some members of the *Laridae* family are birds of the sea while others may be found deep inland in rivers and *jheels*. Some species of gulls are reported to visit land only for breeding. They have long wings and webbed feet. Gulls are expert fliers. The flight is strong and buoyant with slow and often somewhat shallow wing beats. They glide and soar freely and their graceful flight patterns are a delight to watch (16:697). The gulls, larger and stouter than the terns, generally feed on fish and other species of aquatic life. However some are scavengers as well. Their oil glands are well developed with at least three openings on each side (16:697).

Great Black-headed Gull
Larus ichthyaetus
Dhorma in Urdu and Hindi

The Great Black-headed Gull may be found in Pakistan along the coast, in the Indus Delta, on the river Indus right up to Chashma and at the larger Salt Range lakes of Uchhali and Khabakki (2.1:369). It migrates to Central Asia around March-April and is back by October. It is also found on the coasts of India and Sri Lanka.

The Great Black-headed is a very large gull with a wing-span ranging between 1500 mm and 1750 mm (5 feet 8 inches). Its head and neck are black, the back and wings are silvery-grey.

Lesser Black-backed Gull and Great Black-headed Gull can be seen with coots, photograph by Mubashir Hasan

The bill is yellow with a bright red patch at the tip. In winter the head is white, turning black by February. Both sexes look alike.

It feeds mainly on fish and crustaceans but also scavenges and robs other smaller gulls of their food (Roberts' personal comments).

Body length 590-700 mm.
Wing-span 1490-1750 mm.

Brown-headed Gull or Tibetan Gull
Larus brunnicephalus
Dhorma in Urdu and Hindi

The Brown-headed Gull may be found in abundance in Pakistan from September onwards along the seacoast and in the Indus delta and in smaller numbers at all the major canal headworks. Around May it migrates to its breeding grounds in Central Asia, Ladakh and the Tibetan Plateau. In a similar habitat, it is also found in India, Bangladesh and Sri Lanka. It is vagrant in Nepal.

Brown-headed Gull, photograph by Mubashir Hasan

It feeds on fish, prawns and offal, insects, worms and shoots of various crops.

It is a crow-sized bird, grey above and white below. In breeding plumage the head becomes brown but in winter only a darkish patch on the cheek is left. The wing-tips are black but for a short white band on the longer feathers.

Body length 420-460 mm.
Wing-span 1100-1200 mm.

Herring Gull or Lesser Black-backed Gull
Larus argentatus or *Larus fuscus*
Dhorma in Urdu and Hindi

According to Roberts, a bewildering variety of Herring Gulls belonging to the *Larus argentatus/fuscus* group may be found in winter along the coast of Pakistan showing all gradations in the colour of wings and legs. Opinions about their scientific classification vary among scientists (2:1:376).

In size as large as a kite, the Herring Gull, when adult, has a greyish back and wings but the head, neck, under parts and tail are snow-white. Immature birds in their first four years are mottled grey (Roberts' personal communications). It has a bright red patch at the tip of the bill. It is found all along the coast and in the reaches of the rivers lying in-between the large barrages such as Kotri, Sukkur, Guddu, Taunsa, Chashma, Panjnad, and Trimmu. From May to September, it migrates to its breeding grounds of Central Asia around the Black, Azov, Caspian, and Aral seas and the salt lakes of Kazakhstan and Turkestan. It may found in India in a similar habitat.

It feeds on fish, molluscs, crabs, insects, offal and to some extent lives by piracy (1.3:27).

Body length 550-670 mm.
Wing-span 1380-1550 mm.

──── TERNS ────

Family
Sternidae

The terns, more lightly built than gulls, are also expert fliers. Whereas the gulls cannot dive, most terns obtain their food mainly through diving. Their flight is buoyant with faster and deeper wing beats than the gulls'. They rarely if ever glide (17:5). Despite their webbed feet, terns rarely settle on water (1.3:38). Their larger wings and deeply forked tails give them extraordinary manoeuvrability as fliers and divers. They feed on fish and other species of aquatic life. Both sexes look alike.

River Tern or Indian River Tern
Sterna aurantia
Tehari, Ganga Cheel, Machh Louka or *Koorari* in
Urdu and Hindi
Kinai in Sindhi
Krind or *Kreu* in Kashmiri

Found throughout the riverain plains of the Punjab and Sindh the River Tern is a beautiful, slender, pigeon-sized, silvery-grey bird capped with black. Both sexes look alike. It flies endlessly up and down over stretches of rivers, lakes and nullahs, with the most graceful and slow beat of its wings hunting for small fish, crustaceans and aquatic insects. Quarry in sight, it plunges into the water, wings closed, emerging seconds later to resume its hunt.

River Tern, photograph by Mubashir Hasan

Breeding has been reported from the sandbars of the rivers Indus, Jhelum and Chenab and the Nara Canal area (2.1:387). It is absent from Balochistan. It is resident throughout India, Bangladesh, Nepal and Sri Lanka (1.3:49).

Body length 380-440 mm.
Wing-span 800-900 mm.

Black-bellied Tern
Sterna acuticauda
Tehari, Ganga Cheel, Machh Louka or *Koorari* in
Urdu and Hindi
Kinai in Sindhi
Krind or *Kreu* in Kashmiri

The Black-bellied Tern is found throughout the year in all the riverain areas of the Punjab and Sindh. Its breeding sites have been reported along the Indus, Chenab and Jhelum. It is resident, practically throughout India, Nepal, Bangladesh and Sri Lanka.

Black-bellied Tern, photograph by Syed Asad Ali

It feeds mainly on fish, insects and crustaceans and is an exclusively fresh water inhabitant adapted to feed and nest along main river channels (2.1:392). Besides catching beetles and dragon-flies flying over land, it also dives in water like the river tern to hunt for fish.

The black belly of the Black-bellied Tern disappears for a short period after breeding. The front portion of the crown becomes white, the remainder of the crown gets speckled with white spots. In non-breeding plumage the breast and belly is white with some grey. This beautiful, slender, crow-sized tern has long tail feathers which makes its tail appear forked. Both sexes look alike.

Body length 300-350 mm.
Wing-span 700-800 mm.

Whiskered Tern
Chlidonias hybridus
Tehari, Ganga Cheel, Machh Louka or *Koorari* in Urdu and Hindi
Kinai in Sindhi
Krind or *Kreu* in Kashmiri

The Whiskered Tern breeds in Kashmir, northern India and Bangladesh. Although it is found all year round throughout the canal irrigated and riverain areas of the Punjab and Sindh, breeding in Pakistan has not so far been proved. In non-breeding season it spreads over the entire subcontinent.

In breeding plumage, the Whiskered Tern has a narrow white band, alluded to as 'whiskers', which is visible between the jet black cap and black belly of this beautiful pigeon-sized tern. The throat and the belly turn white in winter plumage and the crown becomes grey with dark dots. Both sexes look alike.

The Whiskered Tern feeds on aquatic insects, water beetles, small molluscs and crustacea. Its normal method of hunting is to fly up and down over the surface of land or water, suddenly swoop down to pick its morsel and to resume the hunt. It does not dive into water.

Body length 240-260 mm.
Wing-span 660-710 mm.

Whiskered Tern, photograph by Syed Asad Ali

SKIMMERS

Family
Rynchopidae

Rynchopidae is the family of skimmer's of which only one specie is found in only one valley of the Indus. The birds of this specie are as large as the house-crow with extremely long wings which extend, when at rest, beyond the tail tip which is short and almost square. The head looks large, partly an adaptation to the particular deep, laterally compressed bill. Very specialized feeders, they obtain all their food from the water surface, hunting like other surface feeding terns. Their food comprises a high proportion of small surface-dwelling crustaceans and insect larvae, and when available small fish.

Skimmer, Indian Skimmer or Scissorbill
Rynchops albicollis
Panchira in Urdu and Hindi

In Urdu, *Panchira* means water-shearer, and that is what the Skimmer seems to do, as with consummate skill, it dips the tip of its lower mandible into water, and flies gracefully, skimming the water surface. In the process, as the rushing stream of water gets splashed away on both sides of the bill, small fish are caught in the lower mandible. Down comes the upper mandible catching the prey in the well-fitting grooves of the two mandibles. Promptly, the fish is flipped round and swallowed head first. Singly or in parties of ten or twenty, skimmers slowly

Skimmer, photograph by Mubashir Hasan

beating their wings fly backwards and forwards along the surface of the water as if they are ploughing it (4:488).

The skimmer is rarely encountered away from large rivers (2.1:401). Roberts describes it as a summer breeding visitor to Pakistan although several winter sightings have been recorded. Ali and Ripley describe it as resident and locally migratory, fairly common in the valley of the Indus, northern India and Bangladesh. Roberts considers that the occurrence of the Skimmer in Pakistan is far from common and has declined dramatically with the building of irrigation barrages on the Indus and its tributaries.

The Skimmer is a dark brown and white, crow-sized bird. Above it is blackish-brown, the head, neck and under parts are white, the crown and nape are blackish-brown. Both sexes look alike.

Body length 380-430 mm.
Wing-span 1020-1140 mm.

— SANDGROUSE —

Family
Pteroclididae

Members of this family may be found in desert or semi-desert terrain or in arid regions of low hills, a life for which they are specially suited. They are sandy coloured, pigeon-sized birds with excellent camouflage and are difficult to spot in their natural environment. As in pigeons, plumage is dense with feathers set loosely in the skin. They have short legs and a very short partridge-like neck. Their bodies are compact, elongated and streamlined. The wings are long and pointed. They fly in flocks chattering noisily and their call is a sure clue to identification. The flight is direct and rapid.

Sandgrouse are known for their habit of appearing at a waterhole. At the appointed hour flock after flock will converge from all directions from as far as 80 kilometres (50 miles) away. After a few quick gulps they are off again. In the month of March when some sandgrouse, coming to the waterhole, walked straight into the water and partly submerged their breasts, that action suggested to Roberts that they may have already started breeding as soaking of breasts is regularly used to wet and cool eggs as well as provide water to their chicks (2.1:408).

Chestnut-bellied Sandgrouse, photograph by Syed Asad Ali

Chestnut-bellied Sandgrouse or Indian Sandgrouse
Pterocles exustus
Bhat Teetar in Urdu and Hindi
Batebar or *Bateban* in Sindhi

The Chestnut-bellied Sandgrouse is the commonest sandgrouse of Pakistan and is found throughout the deserts of Thal, Cholistan and Thar all the year round. It is also found in the Sibi plain and along the coastal lands of Makran and Lasbela. It is absent from central Punjab, northern Balochistan and most of the NWFP (2.1:411). It is common in peninsular India (1.3:83).

Both the sexes have a narrow black band across the breast and have pintails. The upper parts of the male are sandy grey and buff with dark narrow marks and yellowish speckles. Cheeks, chin and throat are dull yellow, the belly is chocolate black. The upper parts of the female are dull buff streaked, spotted and barred with dark brown. The upper breast is spotted with black, the lower breast is pale buff, the abdomen and flanks are rufous buff closely barred in black.

It feeds on cultivated grains, weed and grass seeds with which a quantity of grit is swallowed. Breeding season is mainly from March to May but may vary from February to October.

Body length 280-320 mm.
Wing-span 650-700 mm.

Chestnut-bellied Sandgrouse taking off on being disturbed, photograph by Mubashir Hasan

— PIGEONS AND DOVES —

Family
Columbidae

Pigeons and doves belong to the family called *Columbidae*. They are noble birds—peaceful and friendly. The dove is universally recognized as the symbol of peace. Among humans, to be against war is to be a 'dove'.

Fifteen species of pigeons and doves are found in Pakistan (2.1:39). Pigeons tend to be larger than doves. They are grain, seed and fruit eaters. Green pigeons are exclusively fruit eaters. Pigeons feed their young on the regurgitated cellular lining of the crop, a protein-rich cheesy substance known as 'pigeon's milk'. They are strong fliers and are known for their cooing, crooning and whistling voices. Doves often display themselves by flying in a steep climb followed by gliding down in a spiral, sometimes calling.

Blue Rock Pigeon or Rock Dove
Columba livia
Jangli Kabutar or *Kabutar* in Urdu and Hindi
Kentisium in Pashtu
Kapoth or *Chahi* in Balochi
Wan Kotur in Kashmiri

The Blue Rock Pigeon is quite at home in the desolate arid zones of Balochistan, the heights of the Karakoram and in the densely populated areas of metropolises. It is found all over Pakistan except in the desert strips along the southern border of

Blue Rock Pigeon, photograph by Mubashir Hasan

Bahawalpur and Sindh. In India, Bangladesh, Nepal and Sri Lanka, it is quite common.

It is believed to be the ancestor of most domesticated pigeons. Originally it is a cliff dwelling species which has adapted itself to breed in the nooks and corners of buildings, domes of mosques and tombs, ledges of steel and concrete structures, (2.1:416) indeed any place which offers a semblance of a roof overhead and an outlet for it to fly away.

At their favourite routes, flocks of Blue Rock Pigeon can be seen flying to their feeding grounds in the morning and returning from it late in the afternoons. It is a fast flyer and is difficult to shoot except by well practised sportspersons. It is also a prolific breeder, capable of rearing several broods in a year. Both sexes are alike and both share incubation and feeding duties. It feeds on all kinds of grains.

Body length 330 mm.
Wing-span 780-840 mm.

Collared Ring Dove, Indian Ring Dove or Collared Dove
Streptopelia decaocto
Fakhta or *Ghuggi* in Urdu and Hindi
Ghuggi in Punjabi
Gero in Sindhi
Jangli Kapoth in Balochi
Kukil in Kashmiri

The Ring Dove and other members of the family of doves known as *Fakhta* in Urdu and Hindi are generally considered as exceptionally noble birds—peaceful, harmless and trusting. Its deep and sonorous song kukkoo...kook...kukkoo...kook...is interpreted by the village housewife, oppressed by unending monotonous chores of her daily routine, as the story of her life:

Kootoon thi (I was threshing)
Peesoon thi (I was grinding)
Aaya tha (Someone had come)
Gaya tha (Someone had gone)
Kootoon thi (I was threshing [grain])...

Collared Ring Dove, photograph by Mubashir Hasan

The great Punjabi sufi saint and poet Bulleh Shah, lamenting the weakness of the ruler of the day said in one of his well-known lines: *Baz chhad gai alna, Ghuggi pardhan hoi,* (The eagle has forsaken its place and the dove is presiding).

The Ring or Collared Dove is so named for it wears a narrow black half-collar on the back of its neck. It is a pale grey and brown pigeon with a pinkish buff breast. Both sexes look alike. Its range extends from Britain to China. During summer it is found throughout Pakistan except the north-western half of Balochistan and the mountainous areas in the north. During winter it may be seen in abundance in towns as well as the countryside of the Punjab and Sindh (2.1:424). In India, Bangladesh and Sri Lanka, it is quite common.

It feeds on grain and seeds. The nesting season is from May to September.

Body length 300-320 mm.

Red Turtle Dove or Red-collared Dove
Streptopelia tranquebarica
Seroti Fakhta, Girwi Fakhta in Urdu and Hindi

The Red Turtle Dove breeds in the irrigated plains of the Punjab and Sindh. It spends its summer, March to October, in Pakistan and migrates to India where it is resident in most parts of the country and in west Nepal. A few may also be found in lower Sindh the year round.

It feeds on the ground often in the company of the Ring Dove. Like other doves it feeds on grains but weed and grass seeds form a greater proportion of its diet (2.1:426).

The Red Turtle Dove is reddish maroon or purplish chestnut above and below. The head and neck are blue-grey and a short black collar encircles the hind neck. The bill is black and the legs and feet are dull red.

Body length 220-230 mm.

Red Turtle Dove, photograph by Syed Asad Ali

Little Brown Dove or Laughing or Palm Dove (of Africa)
Streptopelia senegalensis
Chhota Fakhta or *Totru* in Urdu and Hindi
Tutan Gheri in Sindhi

Found in abundance in the countryside as well as towns and cities throughout Pakistan, the range of the Little Brown Dove extends from the Philippines to Africa. Its call is pleasant sounding and may be rendered as 'kruu-kukruu-kooh' or 'coo-coo-cuk-rooh' (2.1:438) which W.R.J. Dean calls a quiet, bubbling, cooing call (6:301)—laughing ?

The head and neck of the Little Brown Dove are lilac pink and the rest of the upper plumage is earthy brown with grey patches on wings and shoulders. The breast is pinkish brown and the rest of the under parts are white. Both sexes look alike.

It feeds on grains of wheat, bajra, paddy and lentils besides grass and weed seeds. The breeding season is irregular extending from January to October. The nest is composed of thin twigs, mixed with grass stems and a few roots, nearly meriting Eha's familiar description of a dove's nest as composed of two short sticks and a long one (4:398). According to Stuart Baker 'These

Little Brown Dove, photograph by Mubashir Hasan

birds probably pair for life and are most affectionate to one another. They are also excellent parents and share all duties between them, the hen generally sitting by day when they have eggs, and the cock by night, and the latter also constantly feeds and attends to his wife when she is thus employed.' (18:217).

Body length 260-270 mm.

Spotted Dove or Chinese Dove
Streptopelia chinensis
Totru, Chitte Fakhta, or *Chitroka Fakhta* in Urdu and Hindi

The soft trisyllabic call of the Spotted Dove 'tut-troo-tu or oot-roo-oo' has earned it the name *Totru*. The broad band of black-and-white spots, in a chessboard pattern on the sides and the back of the neck, have earned for this bird the name of the Spotted Dove. It is a pinkish-brown and grey pigeon with a bluish-grey head and neck. Above it is dark brown with pale pinkish spots. The sexes look alike.

Spotted Dove, photograph by Syed Asad Ali

The Spotted Dove breeds in the foothill regions of Punjab between the rivers Jhelum and Indus and of Mardan and Peshawar districts. It may also be sighted in winter in the districts of Lahore and Sialkot (2.1:430). It frequents well-watered and wooded areas, gardens, groves and moist deciduous jungle. It is found throughout India in suitable habitat. It feeds on grains of cereals, grass and weed seeds, and lentils and pulses.

Body length 280-300 mm.

Common Green Pigeon or Yellow-footed Green Pigeon
Treron phoenicoptera
Harial in Urdu and Hindi

In Urdu and Hindi *hara* or *hari* means light green; fresh greenery is called *Hariali* and this beautiful yellowish olive green pigeon has come to be called *Harial*. In the olden days, the call of the partridge and the arrival of the green pigeon symbolized some kind of affirmation of a stable cultural milieu. The writer well remembers, as a child, joining other children in singing on and on just the two lines:

Common Green Pigeon, photograph by Mubashir Hasan

Mamun Teetar bole ga
Harial gadi challe gi

(Uncle partridge will surely call;
The Harial train will surely run)

Its throat, upper breast and mantle are bright olive-yellow and there is a lilac patch near the bend of the wing. The inner surface of the wings is ashy grey. The female looks like the male but its colouring is lighter.

The Green Pigeon is entirely arboreal and is found in flocks in large fruit bearing trees such as pipal and bunyan. On account of the perfect camouflage it wears, once in a tree, it is very difficult to spot. It is expert in clambering around the tiniest of the branches of trees and pecking at the berries. It is uncommon in Sindh, but is found in all the irrigated areas of the Punjab and has not been reported from the NWFP. In India, it is common from the Pakistan border to Bengal, Orissa and Assam.

Body length 330 mm.

PARAKEETS

Family
Psittacidae

The members of the family, *Psittacidae*, included in these pages have grass green bodies, deeply hooked bills and extraordinarily long tails. They are swift fliers and have loud far carrying calls and indulge in much calling when flying around both between feeding places and before setting down in their night time roosts. Largely carnivorous, their hooked bills enable them to eat fruits and seeds that are inaccessible to other birds. They can clamber on tiny branches of small plants and bend their bodies, stretch their necks to pluck, and hold in their feet what they want to consume. For many a farmer and orchard keeper they are a pest of the first order. They can be tamed as pets and taught to repeat words and sentences and perform little tricks.

Rose-ringed Parakeet
Psittacula krameri
Tota, Gallar, Gallar Tota in Urdu and Hindi
Chatun in Sindhi

The Rose-ringed Parakeet is common and widespread throughout the Indus basin and the entire northern part of the subcontinent. It derives its name from the rose pink and black collar around the neck of the male. The body is bright grass green, the bill short, deeply hooked and bright red. It has a pointed tail which is as long as the body itself. The female lacks the colourful collar.

Rose-ringed Parakeet, photograph by Mubashir Hasan

This parakeet feeds on all kind of crops of grain, ripe fruit and on the nectar of flowering trees. Flocks of parakeets have a habit of quietly descending on standing crops and ripening orchards and wreaking havoc unless effectively scared away by watchmen.

The Rose-Ringed Parakeet is a favourite cage-bird in Pakistan, principally for its ability to parrot and mimic. Young birds are assiduously taught words and sentences and are addressed as 'mian mithoo' (dear, sweet one). Once tamed, it does not make a bid for freedom and is let out of the cage which then serves to protect the bird from the predation of lurking cats. It can also be taught to perform a variety of table-top tricks such as loading and firing a toy cannon (1.3:166).

In Pakistan and India, a person capable of sudden changes of allegiance is called *total chashm* (having the eyes of a parakeet). The unfair analogy stems from the sudden change in the colour of the eye of a parakeet. To suit the focusing requirement of a particular moment, as the size of the darker coloured pupil reduces, the lighter coloured iris becomes prominent and in a flash the eye seems to change its colour to yellowish-white from blackish.

Body length 420 mm.

Blossom-headed Parakeet or Plum-headed Parakeet
Psittacula cyanocephala
Tuiyan Tota, Bengali Tota, Desi Tuiyan or *Lal-Sira Tota* in Urdu, Hindi and Punjabi

The Blossom-headed Parakeet frequents well-wooded, moist deciduous plains. In Pakistan, it breeds in the lower Murree foothills and during winter, spreads over the districts of Jhelum and Sialkot. Sightings have also been reported from Islamabad and Lahore (2.1:178). In India it is found along the Himalayan foothills as well as in UP, Gujrat and areas of central India. It

Blossom-headed Parakeet, photograph by T.J. Roberts

feeds on seeds, fruit of all kinds, buds, petals and nectar of flowers.

The Blossom-headed Parakeet is a very colourful bird with a rich purplish-red or plum-coloured head. It wears a black-and-verdigris collar. Its slender body is bright grass green with a small crimson patch on the male's shoulder. The bill is golden-orange on the upper mandible and black on the lower mandible. The central feathers of the long tail are blue and its tip white. The female's head is greyer. It wears no neck ring but has a bright yellow collar.

Body length 330-360 mm.

Slaty-headed Parakeet
Psittacula himalayana
Tuiyan Tota or *Pahari Tuiyan* in Urdu and Hindi
Shoga in Kashmiri

The Slaty-headed Parakeet is found in the valleys of Chitral, Swat, Kaghan and Neelam during summer. Roberts has sighted it in the Margalla Hills near Islamabad during January (2.1: 440). It is a bird of the mountains, not found in the plains. It frequents well-wooded hillsides and valleys, orchards, and neighbourhoods of terraced cultivation. In India it is found in the Himalayas from Garhwal to Bhutan between elevations of 600 metres to 2,500 metres. As the name implies, its head is slaty grey, the body is grass green and the male carries a maroon patch on the shoulder-wing. The bill, short and hooked, is orange-red. The female, slightly smaller in size, is lighter coloured and has no shoulder patch.

The Slaty-headed Parakeet feeds on nuts, acorns, seeds and fruits. Like other parakeets it wastes far more than it consumes and is quite destructive of fruit crops in orchards.

Body length 390-410 mm.

Slaty-headed Parakeet, photograph by Syed Asad Ali

– CUCKOOS AND KOELS –

Family Cuculidae

A loud and persisting, attention demanding or diverting call is known in Urdu as a 'kuk' and the cuckoos and koels are famous for their kuks. In appearance, the members of this family have long and graduated tails with slightly downward curved bills. They are arboreal birds. Some of them are known for making other birds incubate their eggs and bring up the chicks. According to Roberts, recent researches have always shown that the koel extrudes her eggs direct into the nest cavity of the foster parents (personal comments). According to Ali and Ripley, unequivocal data on their breeding biology is lacking.

Pied Crested Cuckoo or Jacobin Cuckoo
Clamator jacobinus
Kala Papeeha or *Chatak* in Urdu and Hindi
Hor Kuk in Kashmiri

The Pied Crested Cuckoo is found in the riverain areas of the valley of the Indus approximately from the end of May to October/November and in India from June to September/October (1.3:195). It is largely absent from the NWFP and Balochistan (2.1:442). It emigrates to Africa for winter. It frequents wooded areas, riverain forests, gardens, and groves in the vicinity of cultivated areas. It is largely an arboreal bird, feeding upon beetles, caterpillars, bugs, ants and termites.

Pied Crested Cuckoo,
photograph by Mubashir Hasan

It breeds throughout the riverain areas of the Indus excepting the desert areas. It is brood parasitic on the Jungle Babbler and the Common Babbler. In October 1983, Roberts observed three adult Jungle Babblers feeding a young fledged cuckoo and a young Jungle Babbler. According to Roberts, since the majority of birds collected from Pakistan are larger in size than those found in south India, the birds found in Pakistan are attributed to the East African migrant population (2.1:443).

It is a myna-sized bird, black above and white below. The prominent bulbul-like crest is also black. Both sexes look alike.

Body length 330-360 mm.
Wing-span 400-500 mm.

Common Hawk Cuckoo or Brainfever Bird
Hierococcyx varius or *Cuculus varius*
Papeeha in Urdu and Hindi

The Common Hawk Cuckoo derives its Urdu name *Papeeha* from its famous call. The bird is immortalized in Urdu and Hindi songs for symbolizing in its impassioned and incessant notes the excruciating distress of the lover torn away from his beloved. '*Pea*' and '*Pia*' in Urdu and Hindi mean beloved. Thus the call

Common Hawk Cuckoo, photograph by Mubashir Hasan

pea-pea-ha-pea-pea-ha

means 'beloved-beloved-come-beloved-beloved-come'. She does not come. Thus the call (rendered by Whistler as pipeeha-pipeeha (4:320), progressively mounts to frantic shrillness and after a dozen or two repetitions breaks off abruptly, to commence all over again after a minute or two. According to Ali and Ripley, on cloudy overcast days and moonlit nights during the peak period, the screaming is almost non-stop (1.3:201).

It is largely an arboreal bird, feeding upon beetles, caterpillars, bugs, ants, termites and occasionally lizards. It is absent from Balochistan and Sindh and much of the NWFP. Its breeding ground is in central and northern Punjab and the valleys of Peshawar and Malakand, where it may be found from March to October. It is widespread all over India, Nepal, Bangladesh and Sri Lanka.

The Hawk Cuckoo is a pigeon-sized bird, above ashy grey, below, white suffused with rufous and ashy on the breast and barred with brownish on the abdomen and flanks (1.3:200). Both sexes look alike. Like other members of the *Cuculidae* family it is brood-parasitic. It leaves its eggs in the nest of Jungle Babblers for incubation and rearing of the chicks.

Body length 330-340 mm.

Koel
Eudynamys scolopacea
Koel in Urdu and Hindi

The two-syllable whistling shriek, 'ko-el-ko-el' or 'ku-oo-ku-oo', increasing in intensity, in ascending scale, with an indefinable quality of excitement, is the well known song of the Koel. The magic of the Koel's kuk has for long filled many poetic breasts with the desire for a reunion with the beloved. Come summer and mangoes, the Koel is there, more to be heard than seen. The irrigated plains of the Punjab and southern Sindh are the Koel's breeding grounds. Except from the extreme south of Sindh, where it is resident, it migrates southwards by October to return in March. In India, the Koel is found in summer in the north and in winter in the south.

Of the size of a crow, it is a bird of groves and gardens, haunting patches of large trees in whose shady boughs it finds concealment and whose fruit it eats. It never descends to the ground (4:327). It feeds on the ripe fruit of bunyan, pipal, ber and mulberry. Occasionally it takes insects, flower nectar, eggs of small birds (2.1:458), and snails (4:327).

The male is glistening black all over, with a yellowish-green bill and crimson eyes. The female is dark brown above, white-spotted and barred white below, spotted on the chin, throat and foreneck, barred in black on the rest of the under parts.

Enmity exists between koels and crows and for good reason. During the breeding season, chiefly May to July but varying locally (1.3:229), the male koel, with strategy and cunning, lures the male and female crows away from their nest. During the short interval of their absence, the female koel either lays or deposits its eggs in the crow's nest destroying one or two of the

Female Koel, photograph by Mubashir Hasan

Male Koel, photograph by Mubashir Hasan

rightful owner's eggs. The crows incubate the eggs and bring up the chicks, their own as well as the koel's. The crow is notorious for its cunning but the koel gets the better of him in this operation.

Body length 400-440 mm.

Common Crow Pheasant or Greater Coucal
Centropus sinensis
Mahoka in Urdu and Hindi

The Crow Pheasant is found the year round throughout the irrigated areas of the Punjab and Sindh and parts of the NWFP. It is absent from Balochistan. It is widespead over the entire northern India.

The Crow Pheasant frequents groves, orchards, sugarcane fields, tall grasses and undergrowth near the edges of rivers and lakes. It is shy and secretive and works its way in and out of thickets. It feeds on lizards, mice, snakes, stranded fish, frogs, insects, molluscs and crustaceans.

Common Crow Pheasant, photograph by Mubashir Hasan

In size, a little larger than a crow, the Crow Pheasant has a glistening black body, head and tail. The wings are bright chestnut.

Body length 480-530 mm.

The garden in Mian Channu where the pictures of Common Hawk Cuckoo, Koels, White-browed Fantail Flycatcher, White-eye were taken.

OWLS

Family
Tytonidae

Owls are birds of prey which normally emerge to hunt well after darkness. They like to spend the daylight hours in the seclusion and concealment offered by the ruins of old forts, deserted habitations, caves, disused wells, as well as in trees with thick foliage providing good hiding places. In Urdu there is a saying: where there are no humans there is the owl.

They feed on small birds, rodents, insects. The hunting technique includes watching from an exposed prominent perch as well as flying low and along hedgerows and ditches where their prey is most likely to be encountered. They are also known to scare roosting sparrows and other small birds at night and to pounce upon them as they fly in confusion trying to settle down once again (author's observation).

Owls are generally brown and white with dark markings. They have large heads and eyes, the face is somewhat flattened giving them binocular vision (Roberts), the legs and feet are feathered. Both sexes look alike.

Barn Owl
Tyto alba
Kuraya or *Karail* in Urdu and Hindi

The Barn Owl is a cosmopolitan species occurring around both hemispheres in temperate and subtropical climates. In Pakistan it is scarce and is confined to the Indus plains (2.1:463). It may

Barn Owl, photograph by T.J. Roberts

be found throughout India, Bangladesh and Sri Lanka. It feeds upon small birds, bats, rats and mice.

The Barn Owl is crow-sized, and is golden buff and grey above, yellowish-brown on shoulders and wings, and white below, spotted with dark brown.

Body length 340-360 mm.
Wing-span 850-930 mm.

OWLS

Family Strigidae

Eagle Owl, Northern Eagle Owl, Rock Eagle Owl
Bubo Bubo
Ghughu in Urdu and Hindi
Gug in Sindhi
Ghubad in Marathi

A large kite-sized solemn bird, according to Whistler (1963, page 342), mottled tawny buff and blackish-brown, with conspicuous tufts above large orange eyes. It is the commonest of the large owls of South Asia. The west of Indus, it is resident

Eagle Owl, photograph by Rolf Passburg

from Chiral to the sea excepting the most arid areas of Balochistan. It is also found throughout Lower Sindh and along the sea coast of Southern Balochistan.

It lives in hollows and clefts of rocky cliffs or ruined buildings, hence the Urdu saying about places where no one resides '*Ulloo bastay hein*'. It is generally a night bird but may be seen even after sunrise perched on the cliff of a high rock.

It feeds on lizards, birds, snakes, rats, mice, and insects and therefore a friend of agricultural economy. It is completely unfair to call it a bird of ill luck for humans. The breeding season extends from December to May.

Body length 580 mm.
Wing-span 400 mm.

Spotted Owlet or Spotted Little Owl
Athene brama
Chughad or *Chakotri* in Urdu and Hindi

The Spotted Owlet has white spots all over its greyish-brown, myna-sized body. Both sexes look alike. It is quite common in

Spotted Owlet, photograph by Mubashir Hasan

the Punjab, Sindh, much of the NWFP and parts of Balochistan as well as throughout India, Nepal and Bangladesh. It feeds on beetles, moths, lizards, mice, earthworms, and small birds.

From a hollow or branch of a tree, as this owlet takes notice, it has a very striking way of staring down into the eyes of the intruder. Without blinking its eyelids, it bobs its head up and down as if it is about to take off for an attack. Roberts believes that in owls, this head bobbing is a threat gesture (personal comments). It is a noisy bird with harsh screechy calls.

Body length 210-230 mm.

Tawny Owl or Himalayan Wood Owl
Strix aluco

The Tawny Owl is a bird of the mountains. In size, it is larger than the House Crow and is found in the Karakorams and the Himalayas from Chitral to Nagaland and Mizoram where it is resident. In the Sulaiman mountain range, it is found in northern Balochistan and northwards along the Afghan border up to

Tawny Owl, photograph by T.J. Roberts

Chitral. In winter it descends to lower altitudes, and has been sighted in the Margalla Hills. It frequents oak, pine, and deodar forests and feeds on birds, lizards, rats and other small mammals. It is common throughout Europe.

The Tawny Owl is mainly a dark brown bird. The crown, mantle and back are mottled and streaked with the feathers being tipped buffy white. The facial disc is whitish. It has no ear tufts.

Body length 450-480 mm.
Wing-span 940-1040 mm.

Short-eared Owl
Asio flammeus

The Short-eared Owl is sparsely distributed in the Punjab and Sindh east of the river Indus and in some areas of Balochistan and the NWFP during September/October to March/April. E. Fernando flushed twenty-two Short-eared Owls in February between the two canals at Marala headworks in northern Punjab

Short-eared Owl, photograph by Tim Hurrell

within an area of about 15 hectares (2.1:490). It is found throughout India, Nepal, Bangladesh and Sri Lanka. It breeds in Asia and Europe from Korea to Scotland.

The Short-eared Owl has two small close-set ears which are not normally visible in the field (2.1:490). It is pigeon-sized, overall pale buff, and heavily streaked with dark brown. The wings and tail are barred rufous and black. Both sexes look alike. It has been seen hunting in broad sunshine (1.3:316). It feeds on beetles, moths, lizards, mice, earthworms, and small birds.

Body length 380 mm.
Wing-span 950-1100 mm.

KINGFISHERS

Family
Alcedinidae

Members of the family *Alcedinidae* are small to medium-sized birds of blue, brown, orange, green, black and white plumage (29:33). Their main diet is fish, large insects and small vertebrates. They prey on fish by diving headlong into the water. They have a large head, compact body, short neck and long, straight and pointed bill. They are perching birds with short and weak legs and feet. As their second and third toes are fused together they perch with difficulty (Roberts' personal comments). On the ground they hop. Both sexes generally look alike (11:413). Their flight is direct and rapid. For nesting they burrow in river banks or adopt holes on trees.

White-breasted Kingfisher, White-throated Kingfisher or Symrna Kingfisher
Halcyon smyrnensis
Kilkila or *Neela Machhrala* in Urdu and Hindi
Dalel in Sindhi
Aspi Chidok in Balochi

The Urdu and Hindi name *Kilkila* for the White-breasted Kingfisher is derived from its call 'killi..li..li...killil..li..li'. It is found right across central and southern Asia from the Philippines to Turkey. In Pakistan it is widespread in the Punjab and Sindh and is absent from drier and higher regions of Balochistan and

western NWFP. It is widely distributed throughout India, Bangladesh, Nepal and Sri Lanka.

It feeds upon small animals, largely insects, ants, termites, small lizards and mice; fish being only a secondary item (1.4:91).

Of the size of a myna, the White-breasted Kingfisher has a large round white patch over its breast. The body and head are deep chocolate brown. The wings, back and tail are iridescent blue. Both sexes look alike.

White-breasted Kingfisher, photograph by Mubashir Hasan

Body length 250-280 mm.

Common Kingfisher, Eurasian Kingfisher or Small Blue Kingfisher
Alcedo atthis
Chhota Kilkila or *Nikka Machhrala* in Urdu and Hindi
Narian Shid in Balochi
Chhota Tuntu in Kashmiri

In Pakistan the Small Blue Kingfisher breeds in the riverain areas of the Indus and its major tributaries, and is more common in Sindh than in the Punjab (2.1:514). It is found in India, Nepal, Bangladesh, Sri Lanka and many other countries of Asia, as well as Australia and Europe.

A little larger than a sparrow, the Small Blue Kingfisher has shiny deep rust-coloured underparts and a brilliant blue back.

Common Kingfisher,
photograph by Mubashir Hasan

Both sexes look alike. It dives in water to prey upon small fish, and also feeds on tadpoles and aquatic insects.

Back view of the Kingfisher

Body length 180 mm.

Pied Kingfisher or Small Pied Kingfisher
Ceryle rudis
Kilkila in Urdu and Hindi
Kingar in Sindhi
Duddru in Kashmiri

The Pied Kingfisher is found throughout the riverain plains of the Punjab and Sindh. It is absent from Balochistan and the desert areas. It is common throughout India, Bangladesh, Nepal and Sri Lanka.

The Pied Kingfisher feeds mainly on fish by diving into the water. To hunt, it flies back and forth, 8-10 metres above the

Pied Kingfisher, photograph by Mubashir Hasan

surface of a river, lake or pond. Espying fish below, suddenly, it checks itself dead in mid-air and hovers with rapidly vibrating wings and positions itself for a strike. It dives headlong into the water with closed wings, as Ali and Ripley call it, a veritable bolt from the blue. Shortly it emerges out of the water, if successful, with the prey in its bill. On a perch the victim is battered and swallowed head first.

Of the size of a myna, the male Pied Kingfisher is white below with a double black gorget across the breast. The female has only a single gorget, broken in the middle. Above, the crown and central part of the crest for both the sexes is black and is finely streaked with white.

Body length 280-320 mm.

BEE-EATERS

Family Meropidae

Bee-eaters are slender, graceful birds with a relatively longer tail, the central part being the longest. The bill is long, slender and down curved. Their principal food consists of bees, giving them their name. Most of their food is obtained by aerial sallies returning to some high up and prominent perch in a tree or on a telegraph wire from which the surrounding air space is surveyed. They are accomplished gliders and flyers.

Little Green Bee-eater
Merops orientalis
Patringa in Urdu and Hindi
Nando Traklo or *Atedan* in Sindhi

The Little Green Bee-eater is found right across Asia, from Africa to Vietnam. In Pakistan, it is absent only from north-western Balochistan and south-western NWFP. It is widespread in India.

The dainty little, paddy-green, sparrow-sized birds, commonly seen launching out from telephone and electricity lines or bush tops, and gracefully gliding back to their perches with a wasp or other insect in their bill, are Little Green Bee-eaters. Being expert fliers, they snap up their prey in mid-air with a dashing upward swoop and whack it against their perch before swallowing it. They feed on bees, wasps, beetles, ants and dragonflies.

Little Green Bee-eater, photograph by Syed Asad Ali

The Little Green Bee-eater is a bright green bird, its head and hindneck are chestnut brown. The chin and throat are blue. The central pair of tail feathers are longer, projecting out like pins. Both sexes look alike. The breeding season extends from February to June with local variations (1.4:110).

Body length 210-230 mm.

Blue-tailed Bee-eater
Merpos philippinus
Bara Patringa in Urdu

The Blue-tailed Bee-eater breeds in Pakistan in a narrow strip extending from Lahore-Balloki to Peshawar-Mardan. It has not been reported from the rest of the country (2.1:523). During the winter it migrates from Pakistan south-eastwards. Its habitat extends up to the Philippines. It is found throughout India, Bangladesh, Nepal and Sri Lanka.

Blue-tailed Bee-eater, photograph by T.J. Roberts

The Blue-tailed Bee-eater feeds on bees, wasps, beetles, ants and dragonflies. It catches its prey in mid-air, takes it to a nearby post and if it is venomous, often rubs or wipes before swallowing (2.1:524).

It is slightly larger than a bulbul, and has a bright green plumage with a bright blue tail and a pale chestnut patch on its throat. Both sexes look alike.

Body length 280-310 mm.

BLUE JAYS

Family
Coraciidae

Rollers and blue jays are members of this family. Compared to the other members of the order *Coraciiformes*, they have comparatively strong, well-developed feet, but their outer toes are fused in their basal part to the central toe, i.e. they are syndactyl. Built stoutly, both sexes look alike with a long-lasting pair bond. They are territorial birds, rather omnivorous in diet and spend most of the day on some perch where they can watch for some prey which consists mainly of lizards and insects or an occasional fish.

Blue Jay or Northern Roller
Coracias benghalensis
Nilkanth, Niltans or *Sabzak* in Urdu and Hindi
Chari in Sindhi
Kangashk in Balochi
Nila Krash in Kashmiri

The arresting sight of the flashing gorgeous blue colours of a Blue Jay in flight is summed up in the Urdu rhyme,

Niltans nila rahio
Meri baat khuda se kahio

(O Blue Jay, for ever, stay blue
And convey my heart's desire to God).

Blue Jay, photograph by Mubashir Hasan

Of the size of a pigeon, the Blue Jay has dark blue and pale blue wings and a rufous brown back, breast and belly. The midcrown is turquoise blue. It is found throughout southern Asia from Iraq to Vietnam minus the desert areas. In Pakistan it breeds in the Punjab, Sindh, most of the NWFP and parts of Balochistan. It is common throughout India.

This opportunistic, omnivorous bird (Roberts' personal comments) feeds upon insects, frogs, lizards and mice and is considered a highly beneficial bird to agriculture, the major proportion in its diet being the insects injurious to agriculture.

Body length 310-350 mm.

HOOPOES

Family Upupidae

The hoopoes are birds, slender in build with a long, thin down-curved black bill adapted for probing in the ground. Its most conspicuous feature is the erectile crest of long narrow feathers down the centre of the crown, like the thatch of a cockatoo. It breeds in Europe, China, Middle East, Africa, the Indo-Chinese region, Malaysia and Indonesia, inhabiting areas which are lightly wooded and preferably where there is some grass covering the ground. A specialized feeder, it is insectivorous and obtains most of its food from beneath the ground surface by probing with its bill.

Hoopoe
Upupa epops
Hudhud in Urdu and Hindi
Katkato in Sindhi
Lachar Ghak in Pushtu
Murgh-i-Suleman in Balochi

Legend has it that, in the wake of the merciless hunting by humans for its golden crest, the Hoopoe petitioned King Solomon to save it from extinction. The mighty king, who understood the language of birds, was pleased to bestow a black-tipped orange-brown erectile crest instead. Thus the Hoopoe was able to survive. In Balochi, it is called Solomon's bird.

Hoopoe, photograph by Mubashir Hasan

Realistic portraits of the Hoopoe have been found in mural paintings of Egypt and Crete. In Western legend the bird is most familiar as the form assumed by Jereus, King of Crete, for his punishment. The Hoopoe is the Lapwing of the Bible. (4:311).

The Hoopoe derives its name from its song which has been rendered as 'hoo-po or hoo-po-po' which is the basis of the English name or hud-hud-hud which is the basis for the name in Urdu. It has a long down-curved bill, the upper part of its body is orange-brown. The back, wings and tail are zebra-like black-and-white. Both sexes look alike.

It breeds in Europe, Asia and northern and southern Africa. In Pakistan its breeding areas are the Punjab and the entire northern and north-western areas of the NWFP and Balochistan. It does not breed in Sindh. According to Roberts, a part of its population in the south migrates to Africa and another part migrates across the Himalayas (2.1:530). In one part of the year or another it is common throughout the subcontinent.

The Hoopoe obtains most of its food by probing into the ground for insects. The young are fed upon caterpillars. Lefroy observed one pair making 286 visits to the nest hole with food over a six hour period (2.1:531).

Body length 290-310 mm.

BARBETS

Family Capitonidae

The members of the family *Capitonidae* are gaudily coloured birds in combinations of reds, greens, yellows and blues with minor patches of black. Their beaks have bristles over the nostrils, hence their name, barbets. With short and rounded wings their flight is seldom long sustained. Except for local movements, they are not migratory (20).

Barbets rarely descend to the ground. They are arboreal birds with a preference for lush green habitat and they feed on fruit, buds and flowers. They climb around tree trunks in woodpecker fashion and excavate nest holes in trunks and branches. Their call is loud and monotonous, often uttered over a long duration from the same perch, but they are never as far away from the hearer as the sound of the call tends to indicate to the human ear.

Blue-throated Barbet
Megalaima asiatica
Bara Basanta in Urdu and Hindi

This is a bird of open hill jungles feeding on a variety of wild fruit and berries, occasionally mantises and other large insects. In Pakistan, the Blue-throated Barbet has been recorded, year round, in the Margalla Hills, Lehtrar, Manga and Kahuta valleys (2.1:536). It is common in Nepal and northern India.

Blue-throated Barbet, photograph by Syed Asad Ali

The Blue-throated Barbet is a large-headed, myna-sized bird, with gorgeous blue throat and cheeks, grass green body, and bright crimson forehead and hindneck. It wears a perfect camouflage and is very difficult to locate in its arboreal habitat. Both sexes look alike.

Body length 220-230 mm.

Coppersmith or Crimson-breasted Barbet
Megalaima haemacephala
Chhota Basanta in Urdu and Hindi

Come spring, come 'tuk..tuk..tuk..tuk', the call of the Coppersmith, at the rate of over a hundred per minute. During winter it is present but mostly silent. Its sighting is rare but around its habitat, one cannot miss the call of this beautiful, sparrow-sized barbet. It has a grass green body, crimson breast and forehead, bright lemon-yellow chin, throat and cheeks, each framed by a black border. Both sexes look alike.

The Coppersmith is found in India, Bangladesh, Sri Lanka, Malaysia, Indonesia, and in other countries of South-east Asia.

Coppersmith, photograph by Mubashir Hasan

In Pakistan, it is a common bird in central and southern Punjab and northern Sindh. It is absent from Balochistan and most of the NWFP as well as from the desert areas of Thal and Bahawalpur. It is an arboreal bird, which feeds upon pipal, banyan and other figs; occasionally on moths.

Body length 140-170 mm.
Wing-span 30-38 mm.

PICULETS,
── WOODPECKERS ──

Family
Picidae

The wood-tapping characteristic has earned the members of the *Picidae* family their general names: *Kathphora* (woodbreaker) and *Thok Barahi* (carpenter tap) in Urdu and woodpecker in English. In proper carpenter-like fashion, they excavate their nest holes in tree trunks. The woodpeckers are associated with trees. They have zygodactyl feet, that is, their two toes point forwards, two backwards, which gives them an excellent grip on the trunks and branches of the trees. They cling to the tree surface vertically facing up but can also suspend themselves on their toes in any other direction. Their movement on a tree trunk is jerky as they creep up in short hurried steps vertically upwards or in spirals. Taking a few rapid spurts of wings at short intervals they fly from tree to tree in undulating curves.

They obtain their food in trees, most of them extracting their prey by tapping the bark, stampeding the insects or finding them in their hiding places. They are largely insectivorous.

Scaly-bellied Green Woodpecker
Picus squamatus
Koel Makots in Kashmiri

Scaly-bellied Green Woodpecker is a pigeon-sized bird with olive-green back, rump and wings. The breast and belly are dull

Scaly-bellied Green Woodpecker, photograph by Syed Asad Ali

greenish grey. The crown is crimson-red in male. Females are similar except that their crown and nape is black. Like all woodpeckers the feet are zygodactyl, that is two facing front and two back.

This woodpecker is a bird of hilly forested areas, occurs from Nepal to north-western Pakistan, in the Salt range, Margalla hills and as far as the forested areas of Ziarat in Balochistan and North-West Frontier. For food it clings to trunks of trees, the stiff pointed tail pressed against the bark functioning as the third leg (Ali and Ripley 1970, vol. 4, p. 183). On the tree trunks it can go up and down with greatest of ease without turning. It feeds on insects, termites, ants, and larvae found on tree trunks.

Body length 350 mm.
Wing length 155-172 mm.

Golden-backed Woodpecker or
Lesser Golden-backed Woodpecker
Dinopium benghalense
Kathphora or *Thok-Barhai* in Urdu and Hindi

The Golden-backed Woodpecker is common in the plains of the Indus and in the valleys of the rivers Kabul and Kurram. It is absent from Balochistan and most of the NWFP. It is widespread in India, Nepal, Bangladesh and Sri Lanka.

The Golden-backed Woodpecker is a very colourful bird; the upper plumage is golden-yellow, the face and the sides of the throat are white; below it is white streaked with black; the male has a crimson crest. The female is similar to the male but its forehead is black speckled with white and only a part of the crest is crimson.

It feeds on insects and larvae, predominantly ants, also on fruits and berries. One was observed clinging to a half ripe mango on a tree, digging into the flesh and swallowing morsels. It takes flower nectar regularly (1.4: 198).

Body length 250-290 mm.

Golden-backed Woodpecker, photograph by Mubashir Hasan

Sindh Pied Woodpecker
Dendrocopos assimilis
Kathphora or *Thok-Barhai* in Urdu and Hindi
Tukok or *Burdi Tokeri* in Balochi
Gihan in Brahui

Sindh Pied Woodpecker,
photograph by T.J. Roberts

The Sindh Pied Woodpecker is found in Balochistan, Sindh, the Punjab and parts of the NWFP. It has not been reported from India. It is a myna-sized, black-and-white woodpecker with white cheeks, crimson crown and vent. The female's crown is black.

It frequents tree-lined canal banks but is also found in desert regions where acacia and thorn trees line the dry seasonal waterbeds. It feeds on insect larvae, pupae and ants (2.1:549). It breeds from March to May (1.4:215).

Body length 220 mm.

Yellow-fronted Pied Woodpecker, Mahratta Pied Woodpecker or Yellow-crowned Pied Woodpecker
Dendrocopos mahrattensis or *Picoides mahrattensis*
Kathphora or *Thok-Barhai* in Urdu and Hindi

The Mahratta Woodpecker is found in India, Bangladesh, Burma, Laos, and Sri Lanka. In Pakistan it is found throughout the riverain areas of the Punjab and Sindh. It is absent from Balochistan. It feeds on bark-dwelling insects and larvae, moths, ants, termites, caterpillars, but is adaptable in exploiting flower nectar, fruit and berries (2.1:554).

The male Mahratta Woodpecker is a bulbul-sized woodpecker, brownish-black above; its forehead is brownish-yellow and the back part of the head is scarlet. The chin and throat are white and the rest of the under parts are streaked with brown. The female is similar to the male but its crown is golden-brown with no scarlet in it.

Body length 170-180 mm.

Yellow-fronted Pied Woodpecker, photograph by Robert A. Randall

LARKS

Family Alaudidae

The members of the *Alaudidae* family are ground-dwelling birds of small size. Both sexes look alike. The colours of earth dominate their plumage giving them almost perfect camouflage which makes them difficult to spot on the ground. But they are great songsters. The melodious arrangement of their tunes attracts attention. The bird is nowhere in sight. Where on earth is the song coming from, the hearer wonders? It is coming from up above. The lark is flying and singing. The song and the display of flight is fascinating.

Most larks live in open country, on grassy plains, cultivated fields, deserts or beaches. All larks have pointed, slightly down-curved bills and most have crests or tufted heads (20). They feed on insects and seeds. They walk or run but do not hop.

Red-winged Bush Lark
Mirafra erythroptera
Aggia in Urdu and Hindi

All of a sudden one hears the rather simple song of the male Red-winged Bush Lark and spots it rising vertically up from the ground. Up it goes for about ten metres, the tempo of the song slows down and fades. The bird is seen floating or 'parachuting' down, wings held motionless and upstretched, still singing. The entire performance lasts for about twenty seconds (1.5:8 and 6).

Red-winged Bush lark, photograph by Syed Asad Ali

According to Roberts who has tape-recorded songs of a large number of the birds of Pakistan: amongst the larks, the Bush-larks are the poorest songsters (personal communication).

It feeds on grass, weed seeds and insects and is found only in southern Sindh (2.2:6). Ali and Ripley report it to be common in north as well as south India, except Kerala. Its breeding season lasts from April to September.

The Red-winged Bush Lark is a sparrow-sized bird with prominent chestnut in its wings; above it is fulvous brown streaked with blackish; the chin and throat are whitish.

Body length 140 mm.

Black-crowned Finch-Lark
Eremopterix nigriceps
Duri or *Dabak Chiri* in Urdu and Hindi

Having a preference for arid regions, the Black-crowned Finch-Lark is found in southern Balochistan, parts of Sindh and in the deserts of Thal and Cholistan (2.2:7). It is also found in

Black-crowned Finch-Lark, photograph by Syed Asad Ali

the countries of south-west Asia and north-east Africa and parts of India (Ali and Ripley).

The upper plumage of the female and its tail are dark brown tinged with grey and rufous, the lower plumage is pale rufous. The male has a chocolate brown crown and is ashy-brown above with a large whitish patch over the cheeks. Below, the plumage is dark chocolate brown.

It feeds on grass, weed seeds and insects. Its breeding season is irregular—between February and September.

Body length 130 mm.

Ashy-crowned Finch-Lark
Eremopterix grisea
Deoli or *Dabak Chiri* in Urdu or Hindi

The Ashy-crowned Finch-Lark is found throughout the Indus plains, and in India, Nepal and parts of Sri Lanka. It is a sparrow-sized bird of open country, away from trees in fallow

Ashy-crowned Finch-Lark, photograph by Syed Asad Ali

and ploughed fields, scrub ground or wasteland where it is difficult to spot on account of its near perfect camouflage. It feeds on grass and weed seeds, ants and other insects.

During the breeding season which in Pakistan is mainly during and after the monsoon, the male sings all the time on the ground as well as in the air. The song is sweet but monotonous 'treedle-dlee-dlee', without variation (4:263 and Roberts' personal comments).

The upper plumage of the female and its tail are dark brown tinged with grey and rufous, the lower plumage is pale rufous. The male is ashy-brown above with a large whitish patch over the ears. Below, the plumage is dark chocolate brown (4:262).

Body length 130-135 mm.

Greater Short-toed Lark or Yarkand Short-toed Lark
Callandrella brachydactyla
Pullak or *Bagheri* in Urdu and Hindi

There seem to be problems of classification with the genus *Calandrella*. Ali and Ripley classify the Yarkand Short-toed as *Calandrella cinerea longipennis* whereas according to Roberts, the Greater Short-toed is a species distinct from *Calandrella cinerea*. Ali and Ripley and Roberts both consider what they call the Greater Short-toed Lark to be common in the valley of the Indus. According to Roberts, the race he calls the Rufous Short-toed Lark is a bird that migrates between Central Asia and south India, and only passes through Pakistan and is out of his checklist of *The Birds of Pakistan*. Be that as it may, the Short-toed Lark of Ali and Ripley and the Greater Short-toed Lark of Roberts, is a sparrow-sized bird found throughout Pakistan, except western and central Balochistan and parts of the NWFP. For breeding it migrates to Central Asia at the end of February to April to return in September/October. It is widespread all over India. During migration, the size of the

Greater Short-toed Lark, photograph by Mubashir Hasan

flocks can be in the thousands. They roost at night on bare open flats and sometimes fly out great distances every morning to drink. They feed on grass and weed seeds and insects.

This lark is sandy or greyish-brown above streaked with blackish. Below it is white, the breast finely striated with brown. Both sexes look alike.

Body length 150-165 mm.

Crested Lark
Galerida cristata
Chandul in Urdu and Hindi

The Crested Lark is a bird of the open country and is found throughout Pakistan and much of India. It lives on the ground and prefers country tracks and roadsides where it may be seen running on the ground picking grain and weed seeds and chasing insects. It freely perches on bushes, wires and fence-posts. It is spread widely over eastern Europe, northern Africa, and west Asia (4:259).

Crested Lark, photograph by Mubashir Hasan

Larger than a sparrow but smaller than a bulbul, the Crested Lark has an upstanding pointed crest as long as its bill. Above it is sandy brown with blackish streaks. Below it is white but streaked with brown on its breast.

Its song, a sweet and plaintive 'ti-ee' or 'tee-urr' is delivered during flight as well as sitting. Its breeding season is March to August, mainly April to June (1.5:38).

Body length 170-180 mm.

Hoopoe Lark, Persian Desert Lark or Bifasciated Lark
Alaemon alaudipes
Rann Candle in Urdu, Hindi and Gujrati

The Hoopoe Lark is one of the largest and most distinctive of the larks. As big as a myna, it is confined to pure desert areas and is thinly scattered in Balochistan, Sindh and southern Punjab in Pakistan and the Great and Little Rann of Kutch in India. It is also found in north Africa, Arabia, Iraq, Iran, and Afghanistan.

Hoopoe Lark, photograph by Khan Mohammad

The Hopoe Lark is known for its prolonged musical whistling and piping song which is uttered on the ground or during display flight. It is a very swift runner, moving about in the desert, suddenly stopping to pull itself erect or to pick up a morsel. It feeds upon seeds, tiny beetles and other insects. (1.5:17-18)

Its upper plumage is sandy grey and the lower plumage is whitish and streaked black on the breast. Its black bill is curved and legs are white. Sexes look alike, the female being a little smaller than the male. Breeding season lasts from March to July. According to Roberts, it builds its nest on the top of a Suaeda bush or clump to prevent the eggs being buried by blowing sand (personal comments).

Body length 230 mms.

— SWALLOWS, MARTINS —

Family
Hirundinidae

The members of the *Hirundinidae* family are known for their graceful flight and regularity of migrations. In Korea and Japan a swallow nesting on one's house is considered a sign of good luck. They are small, slender, long winged birds with a short neck and forked tail. Their tiny triangular bill with a wide gap has bristles which act as aerial fly scoops (20:216). They are highly efficient fliers with great capability of turning and gliding. Swallows spend much of their time on the wing, hawking back and forth for insects.

Despite the known habit of swallows for returning to their regular nesting places, many cases have been recorded of swallows nesting on moving vehicles—boats, trains and tractors. They seem to have no problems keeping up with the movement of the vehicle. A pair of Grey-breasted Martins stayed faithfully with a boat and reared brood after brood as the steamer went its weekly rounds up and down the Essequibo River, a round trip of 180 miles (20:218). Their plumage is commonly dark blue, glossy above and paler below. They are found all over the world except the polar regions and have a preference for open country and proximity of fresh water. They feed on insects caught on the wing.

Brown-throated Sand Martin or Plain Sand Martin
Riparia paludicola
Ababil or *Abali* in Urdu and Hindi
Ababil Paki in Sindhi

The Plain Sand Martin is a sparrow-sized bird, grey-brown above and white below with a diffuse grey band across the breast. Both sexes look alike. It is found throughout the valley of the Indus and along the Makran coast (2.2:31). It is seen perched in closely packed rows on telegraph and electricity wires. It is common throughout India, Nepal, Bangladesh and Sri Lanka.

The Plain Sand Martin nests in a tunnel, often a metre or more deep which it makes in sandy cliffs and banks. A nesting colony of a hundred tunnels or more, honeycombing a steep sandy bank of a river or an embankment forming the edge of a lake, is not unusual. It feeds on insects which are taken on the wing.

The name *Ababil* in Urdu and Hindi is taken from Arabic. This name is mentioned in the holy book of Islam in referring to a pre-Islamic episode in which flocks of *Ababils*, by throwing

Brown-throated Sand Martin, photograph by Mubashir Hasan

pebbles, forced an army employing elephants to abandon its invasion of the shrine at Mecca.

Body length 120-130 mm.

Wire-tailed Swallow
Hirundo smithii
Leishra in Urdu and Hindi

The Wire-tailed Swallow has wire-like streamers jutting out of the pair of its tail feathers, the length of the wire being longer than the body of the swallow. It is a resident of southern Sindh and is a summer breeding visitor to the valley of the Indus in Pakistan (2.2:40). It is common in Bangladesh, Nepal and India. It is rarely away from canals, rivers or other bodies of water.

The Wire-tailed Swallow is an expert flier. Not only does it prey upon insects on the wing, but also, once the young are able to fly, they are fed in the air. The parent and the youngsters circle round and round and then for an instant cling together

Wire-tailed Swallow, photograph by Rolf Passburg

during which the mouthful of insect is transferred to the youngster (4:239).

The upper plumage, sides of the head and neck are glossy steel blue, the top of the head is bright chestnut and the lower plumage is white. Both sexes look alike except that the 'wire' is shorter in the female.

Body length 140 mm.
Streamers 175 mm. extra.

Red-rumped Swallow or Striated Swallow
Hirundo daurica
Masjid Ababil in Urdu and Hindi
Phairni in Kashmiri

The Red-rumped, Striated or Mosque Swallows are found right across Eurasia from southern Europe and Africa to China including Pakistan and India. According to Roberts, there is no resident population of this swallow in Pakistan, although stray individuals may be encounterd in the southern parts of Sindh. It

Red-rumped Swallow, photograph by Syed Asad Ali

is mainly a summer breeding visitor to Balochistan, the NWFP and northern areas of the Indus valley (2.2:42).

Its favourite spots for nest making are the ceilings or domes of ancient mosques, tombs, under bridges or natural rock overhangs (1.5:71). It is a great flyer, spending the hours of daylight on the wing, preying on insects, not so much above the surface of water as other swallows but on drier country.

The Red-rumped Swallow has a chestnut rump and nape; above it is glistening steely black and below fulvous white, finely streaked with dark brown. It is a little larger than the sparrow in size. Both sexes look alike.

Body length 190 mm.

Cliff Swallow
Hirundo fluvicola
Nahar Ababil in Urdu and Hindi

The Cliff Swallow derives its English name from its habit of building its nest on the face of overhanging cliffs and its Urdu

Cliff Swallow, photograph by Mubashir Hasan

name for building its nest beneath the arches or slabs of bridges over canals. It is found throughout the Indus plains except in the southern parts of Sindh. It is common in India.

Whistler describes this swallow as one of the 'purely social swallows' (4:240) as it spends all its life in big flocks which never separate. It nests in colonies, often in clusters of a hundred or more.

In size, the Cliff Swallow is smaller than the sparrow. Above, it is glossy steel blue with a dull chestnut forehead and crown and a pale brown rump. Below it is fulvous white, streaked with blackish on the sides of the head, throat and breast. Both sexes look alike.

Body length 120 mm.

── PIPITS, WAGTAILS ──

Family
Motacillidae

The members of the family *Motacillidae* are slim and long tailed, sparrow-sized birds of black, grey, yellow, olive and brown plumage. The pipits are brown and olive and the wagtails are a combination of black, white, grey and yellow. While the pipits are shy and alert, the wagtails are cheerful and confiding, constantly showing their friendly nature by wagging their tail.

Scientists believe the yellow-headed and grey-headed races of the Yellow Wagtails come from the same parental stock. They got isolated from each other during the Pleistocene Ice Age as the ice sheet contracted. The grey-headed moved northwards into Europe while the yellow-headed got split into two groups, one in Britain and the other in Asia Minor (Tree. A.J., *Complete Book of South African Birds*, page 580).

More than a dozen races of wagtails are classified under four species, namely: Yellow Wagtails, *Motacilla flava*, Yellow-headed or Citrine Wagtail, *Motacilla citreola*, Grey Wagtail, *Motacilla cinerea*, and White and Pied Wagtails, *Motacilla alba*. Their exact identification, especially of the Yellow Wagtails, when in winter plumage, presents considerable difficulty. Some females and juveniles are almost impossible to identify correctly. The wagtails are not found in arid areas.

The pipits and wagtails walk or run and have undulating flight. They have a longish tail which is constantly wagged vertically. They feed on tiny insects and other invertebrates, living near or in the grass. In Manipur, eastern India, all wagtails are regarded as an incarnation of the goddess Durga.

Paddy-field Pipit or Richard's Pipit
Anthus novaeseelandiae
Charchai or *Rugail* in Urdu and Hindi

The Paddy-field Pipit is a sparrow-sized bird frequenting grassland, edges of cultivated fields, and grazing grounds. It is found throughout Pakistan, India and Bangladesh except in Balochistan, western parts of Sindh and the NWFP and in the mountainous areas. It has also been reported breeding in some places in the Himalayas and other ranges up to about 1,500 metres. It feeds on bugs, spiders, weed seeds, grass blades and other vegetable matter and insects.

The Paddy-field Pipit is usually found in pairs, briskly running on the ground and moving its tail up and down. It is a fulvous brown bird, pale fulvous below and streaked on the breast, the wings and tail are dark brown and so are the sides of the throat and foreneck.

Body length 150 mm.

Paddy-field Pipit, photograph by Mubashir Hasan

Tawny Pipit, photograph by Mubashir Hasan

Tawny Pipit
Anthus campestris
Chillu in Urdu and Hindi
Dhan Chidi in Gujarati

The Tawny Pipit, a sparrow-sized bird, is found all over Pakistan and a greater part of the Indian Peninsula and east to the Brahmaputra river in Bangladesh (1.9:256). Around the middle of April, it leaves for its breeding grounds which extend from Spain to Mongolia. It also breeds in Morocco, Turkestan and north-eastern Afghanistan. The return migration takes place during September.

The Tawny Pipit frequents semidesert, fallow or ploughed fields, pastures, and sparsely scrubbed stony country. It feeds on insects and weed seeds. It is pale brown, lightly streaked above. It is plain whitish buff below, sometimes with dark streaks on the breast. Both sexes look alike.

Body length 150-160 mm.

Brown Rock Pipit, photograph by Mubashir Hasan

Brown Rock Pipit, Long-billed Pipit or Persian Rock Pipit
Anthus similis

The Brown Rock Pipit is a large bulbul-sized bird which despite its name, does not have a significantly larger bill than other similar pipits. Roberts records the occurrence of this pipit mainly west of the river Indus, the Punjab Salt Range and adjacent areas being the exception. It is also reported to be absent from southern and western Balochistan. It breeds between 600 and 1,800 metres, in Balochistan, the NWFP, Kashmir, the Karakorams and western Himalayas. It spreads out during winter though still remaining within the confines mentioned earlier. The Brown Rock Pipit is also found in Iran, Palestine, Lebanon and southern Africa. Typically it frequents semidesert terrain in rocky areas with sparsely scrubbed and stony slopes. It feeds on insects and berries.

The Brown Rock Pipit is pale brown above, slightly streaked on the head and back. The wings and tail are darker. Below, the throat is whitish and the rest of the underparts are pinkish buff with faint brown streaks, sometimes wanting. Both sexes look alike.

Body length 200-230 mm.

Yellow Wagtail
Motacilla flava
Dhoban, *Pila Mamola* or *Pilkya* in Urdu and Hindi

Four races of the Yellow Wagtails occur in the valley of the Indus: Blue-headed, Grey-headed, Black-headed and White-headed. This species occurs in a number of distinctively patterned population (Roberts' personal comments). From September/October to April, these are abundant in the Punjab, Sindh and parts of the NWFP and Balochistan. They are also extensively found in India, Nepal and Bangladesh. Migration to their breeding grounds commences in March and continues till the end of May.

The wagtails feed on insects including bugs, beetles, flies, and weevils. In the cities, a lawn under irrigation invariably

Yellow Wagtail, photograph by Mubashir Hasan

attracts a party of wagtails. Along the edges of lakes, in moist pastures or grasslands, on canal banks, among grazing cattle or in the company of washermen working on the edge of a pond, wagtails are sure to be found. Their faithful attendance at the washerman's work place has earned them the name of *Dhoban* in Urdu and Hindi.

The Blue-headed Wagtail is a sparrow-sized bird and has a pale bluish-grey head. The back is olive, and the wings are brownish. Its chin, breast and belly are bright yellow. Both sexes look alike.

The Grey-headed has a dark slaty grey head and nape. It has brown wings with two yellow bars. The head of the Black-headed is black. In other respects it is similar to the Grey-headed.

Body length 170-180 mm.

Yellow-headed Wagtail
Motacilla citreola
Pila Mamola, *Pilkya* or *Pani-ka Pilkya* in Urdu and Hindi

Yellow-headed Wagtail, photograph by Mubashir Hasan

The Yellow-headed Wagtail, a sparrow-sized bird, is found in winter in the Punjab, Sindh and parts of the NWFP and Balochistan. It is also found in India, Nepal and Bangladesh. Northward migration to its breeding grounds in alpine regions of Pakistan commences in March and continues till the end of May. Like other wagtails it feeds on insects, bugs, beetles, and tiny molluscs. It frequents edges of lakes, in moist pastures or grasslands and would also look for food on the leaves floating on the surface of water such as those of the lotus.

The male of the Yellow-headed Wagtail has a rich lemon-yellow head, breast and belly; the back and rump are black; wings and tail are dark brown. The female is variable. It is either like the male or paler; some individuals may have a dark greyish crown and mantle.

Body length 160-177.5 mm.

Grey Wagtail
Motacilla cinerea or *Motacilla caspica*
Dhoban or *Dhoban Chirya* in Urdu and Hindi
Balkatara in Punjabi
Khak Dobbai in Kashmiri

Like other wagtails, the Grey Wagtail is sparrow-sized and is very widely distributed in Pakistan, India, Nepal and Bangladesh. It breeds between elevations of 1,200 to 4,000 metres in the mountain ranges of the Sulaiman, the Hindukush, the Karakorams and the Himalayas. Its winter population is augmented by the flocks breeding in Central Asia and further north. It feeds on insects, butterflies, and tiny molluscs. For its faithful attendance at the washerman's workplace it has earned the name of *Dhoban* in Urdu and Hindi. It is abundantly found in the cities where lawns are being irrigated, along the edges of lakes, in moist pastures or grasslands, and on canal banks among grazing cattle.

Grey Wagtail, photograph by Mubashir Hasan

The Grey Wagtail has a grey head and back, dark brown wings, a blackish-brown tail, black throat and in breeding plumage, the rest of the underparts are yellow. The wing feathers have whitish edges and so has the black of the throat.

Body length 180-200 mm.

White Wagtail
Motacilla alba
Dhoban or *Dhoban Chirya* in Urdu and Hindi
Balkatara in Punjabi
Peenchkani or *Dobbai* in Kashmiri
Kahtriani (woman dyer) in Gujrati

The White Wagtail is perhaps the commonest of the wagtails in urban areas as well as in the countryside. It is partial to the neighbourhood of water. In the cities it is present in all the moist or underwater lawns of the parks. Along the edges of lakes, in moist pastures or grasslands, on canal banks, among grazing cattle or in the company of washermen working on the edge of a pond, wagtails are sure to be found. Its faithful attendance at the washerman's workplace has earned it the name of *Dhoban* in Urdu and Hindi. It feeds mainly on insects; larvae, flies, weevils, beetles and caterpillars are also taken. It is found throughout Pakistan, India, Nepal, Sri Lanka and Bangladesh. It breeds, during summer, in the Karakorams and the Himalayas and further north in Asia and Europe from Japan to Britain.

For a bird of its size, the White Wagtail has the longest tail which constantly wags up and down. Its upper plumage is ashy grey, wings and tail are black, the feathers broadly margined with white. In winter plumage except for a black crescent-like collar on the breast, the remainder of the head and lower plumage are white, tinged with ashy on the flanks.

Body length 180-190 mm.

White Wagtail, photograph by Mubashir Hasan

Large Pied Wagtail
Motacilla maderaspatensis
Mamula, Bhuin Mamula or *Khanjan* in Urdu and Hindi

The Large Pied Wagtail is a bulbul-sized bird found the year round, in the northern riverain areas of the valley of the Indus. It is also found in India and parts of Bangladesh. It feeds on beetles, locusts, dragonflies, snails and small seeds. It frequents edges of ponds, river banks and embankments, smooth running streams and occasionally rice paddies and lawns.

The upper plumage of the Large Pied Wagtail and its breast, wings and throat are black with a large white band on the wing; the lower plumage is white but ashy on the flanks. Both sexes look alike.

Body length 210-240 mm.

Large Pied Wagtail, photograph by Mubashir Hasan

MINIVETS

Family
Campephagidae

Campephagidae is the family of cuckoo snikes, wood snikes and minivets. It is a large family inhabiting the Himalayas and peninsular India on lower altitudes. The minivets are usually found all over the Punjab, and in some areas of Sindh and Balochistan. These birds vary in colour with body size ranging between 160 to 230 mm. They are a noisy family whose birds call throughout the day all year long.

Long-tailed Minivet
Pericrocotus ethologus
Pahari Bulal-Chashm in Urdu and Hindi
Wozul Mini in Kashmiri

The Long-tailed Minivet is a bulbul-sized bird. The male is a glossy black and scarlet bird which breeds (April to June) in northern parts of the NWFP, Gilgit, Kashmir and westward up to Nepal and winters from October to March in the plains of the Indus and northern India. It is a strictly arboreal bird found in Himalayan forests, and in winters in mango groves, orchards, gardens and trees bordering cultivated lands. It feeds on spiders, beetles and other insects and buds of acacia and fruits.

The male of the Long-tailed Minivet has a glossy black head, throat and the wings have a large scarlet patch; the rest of the body is deep scarlet. The black and scarlet tail is graduated. In

Long-tailed Minivet, photograph by Syed Asad Ali

the female the black areas are replaced by grey and the scarlet by yellow.

Body length 180-200 mm.

Small Minivet or Wandering Minivet
Pericrocotus cinnamomeus
Bulal-Chashm, *Rajalal* or *Saheli* in Urdu and Hindi

The Small Minivet is a sparrow-sized bird of semi-desert environment in the vicinity of cultivated areas endowed with clumps of large trees. It feeds on moth caterpillars and other insects and is found throughout the irrigated plains of the Indus and its tributaries. It is absent from the mountainous and arid areas but does occur in Sindh Kohistan which is both hilly and arid. It is common throughout India, Nepal and Sri Lanka.

The throat, head and back of the male Small Minivet is dark grey; the wings are black with a yellow-orange patch; the rump is orange-red; the breast is bright orange becoming yellow on

Small Minivet, photograph by Syed Asad Ali

the belly. In the female, the upper parts are as in the male but paler; the under parts are whitish grey and the throat is not black. With slight variations in its colour plumage, the Small Minivet is found throughout India and parts of Nepal and Burma.

Body length 150 mm.

BULBULS

Family Pycnonotidae

The members of the *Pycnonotidae* family are found in much of Asia and Africa: Iraq, Iran, Afghanistan, in the countries of South Asia, South-East Asia, in China, Japan and in north and south Africa. The *Pycnonotus jocosus*, the Red-whiskered Bulbul, has been successfully introduced in Florida (8:486). There are 21 species of bulbul altogether inhabiting the subcontinent and of these four have red or yellow patches under their tail (Roberts' personal comments). The bulbuls common in the Indus plains are soberly coloured birds, often with a crimson or yellow patch under the tail. Their unremarkable appearance, however, is only offset by the prominent place they have won among the people as one of the best known birds. In Persian and Urdu poetry, the bulbul occupies a place of high honour as a lover, a songster, a wailer, and an innocent quarry of the netter of birds.

The bulbul is known for the obvious affection it has for its mate, the two never being far away from each other, are often seen snuggled together on a perch. In poetry, however, it is admired more for its assumed exemplary love of the flower, the poets preferring not to take notice of the fact that the nectar and the buds of flowers are items of its food.

There is an enchanting light-heartedness in the voice of the bulbul. Although it has no song of its own as such, it has a repertoire of a few highly melodious notes. Its calls greatly add to the pleasure of enjoying a garden or a park. One likes to hear its 'Be care-ful', 'Pleased to see you' or 'Tea for two', or just 'Be quick-quick' calls again and again, without getting tired of them. However, the bulbul, the songster of Persian literature, is quite another bird; *Erithacus megarhynchos hafizi*, called bulbul in Persian and nightingale in English. The latter is occasionally sighted in Balochistan.

Bulbuls are aggressive and quarrelsome birds and courageous fighters and as such are netted and sold to be tamed as fighting pets.

White-cheeked Bulbul
Pycnonotus leucogenys
Bulbul in Urdu and Hindi
Kushandra in Punjabi
Bhooroo in Sindhi

The White-cheeked Bulbul is found in all the provinces of Pakistan and in the north-western states of India. It is a bird of semidesert tracts, scrubs, groves, gardens and cultivated areas. It is a crestless grey-brown bulbul with a black head and throat and a large white cheek patch. There is a yellow patch under the tail. The under parts are white. Both sexes look alike. It feeds

White-cheeked Bulbul, photograph by Syed Asad Ali

on berries, seeds, flower buds, nectar, caterpillars, ants and other insects. Its general habits are the same as described earlier under Bulbuls.

Body length 200-220 mm.

Red-vented Bulbul
Pycnonotus cafer
Bulbul, Kala Bulbul, Thar Bulbul, Gul-Dum in Urdu and Hindi
Bhooroo in Sindhi

The most common of the bulbuls in the valley of the Indus, the Red-vented is found wherever there are gardens, light forests, shrubbery in bungalows, light scrub and evergreen patches with trees. Except for some areas of Balochistan and the NWFP, it is found all over Pakistan, India, Nepal, Bangladesh and Sri Lanka.

It feeds on berries, seeds, flower buds, nectar, caterpillars, ants and other insects. Its general habits are the same as described earlier under Bulbuls.

Body length 200-230 mm.

Red-vented Bulbul, photograph by Mubashir Hasan

Black Bulbul, photograph by Syed Asad Ali

Black Bulbul or Himalayan Black Bulbul
Hypsipetes madagascariensis
Pahari Bulbul in Urdu and Hindi
Wan Bulbul or *Kruhun* Bulbul in Kashmiri

The Black Bulbul is ashy-grey throughout with a bright red bill, legs and feet. It has a black crest and a slightly forked tail. It is a bird of high forest trees and slightly larger in size than other members of the family. During summer it is found in the valleys of Chitral, Kaghan, Kashmir and further east in the Himalayas up to Nepal, Sikkim and Bhutan. It descends in winter to the foothills in the west and the plains in the east but has been recorded as a straggler in Kohat and Lahore. It feeds on berries, seeds, flower buds, nectar, caterpillars, ants and other insects. It keeps in parties of six to ten, but sometimes numbering a hundred (1.6:1148). It is one of the finest fliers among the members of the bulbul family.

Body length 240-254 mm.

SHRIKES

Family Bombycillidae

Grey Hypocolius or Shrike-Bulbul
Hypocolius ampelinus

The Grey Hypocolius is a true desert dweller confined to the southern parts of south-west Asia and North Africa and until recently was considered a rare vagrant into Sindh and Balochistan, in the Indian Rann of Kutch and Maharashtra. Apparently not so any more. Roberts describes it as an irregular but not uncommon visitor to the remote desert tracts of southern Balochistan. Notable among recent sightings just west of Karachi

Grey Hypocolius, photograph by Syed Asad Ali

in the Hub valley, are by Roberts, Passburg, and Asad Ali in 1984 to 1986, and 1989.

The Hypocolius is a bulbul-sized bird, resembling a Long-tailed Grey Shrike, as also the non-Asiatic race of the White-eared Bulbul. It is blue-grey above and pinkish below. The female is greyer above and more creamish below and lacks the bold black marking on the cheeks. It is an arboreal bird, which feeds on figs, dates, berries; essentially a fruit eater but will help itself on insects as well.

Body length 230-250 mm.

ACCENTORS OR 'HEDGE SPARROWS'

Family
Prunellidae

The members of the family *Prunellidae* are small sparrow-like creatures of mountainous regions with plumage black, brown, grey or buff, streaked above and plain or streaked below. Typical of the accentors, they forage on the ground, with a rather low crouching gait, hopping over rather exposed open places. Rather tame, their flight is strong and direct when disturbed. Their food is mainly insects during the summer and consists of fallen seeds in the winter, supplemented with berries.

Alpine Accentor
Prunella collaris

The Alpine Accentor is a sparrow-sized bird of the mountains. It breeds between elevations of 3,600 and 5,000 metres in stony slopes, cliffs and morains above the timberline and descends to 1,800 metres during the winter. It is found in Safed Koh, Chitral, Gilgit, Baltistan, Kashmir, Ladakh and the Himalayas up to Sikkim and Bhutan. Also found in Afghanistan and Turkestan (1.9:145).

The Alpine Accentor feeds on insects and small seeds. Its head is greyish-brown; the back is streaked with greyish-brown and the tail is dark brown. The chin and the centre of the throat are white, finely barred with brown; sides of the throat, breast

Alpine Accentor, photograph by T.J. Roberts

and centre of the belly are grey and the flanks are rusty. Both sexes look alike. Its song is almost skylark-like (5:220).

Body length 155-170 mm.

– THRUSHES AND CHATS –

Family
Turdidae

The family *Turdidae* comprises thrushes, chats, wheatears, robins, redstarts, nightingales, and some others. They can justifiably claim to have among them some of the finest songsters of the bird world. Most of the smaller species are known as chats. They are slender billed, strong legged song birds of medium size and they are arboreal or terrestrial. They are found on almost any kind of terrestrial habitat, from arid desert to arctic tundra to dense equatorial forest (19:592). Thrushes and chats are monogamous and maintain pair bonds in the breeding season and exclusive breeding territories.

Robins and nightingales are found all over northern Europe and Asia, south-east and south Asia and in Africa. Magpie robins and shamas are members of this group. Although the European Robin shuns open country (Roberts' personal comments) many species in the valley of the Indus are open country chats such as stonechats, redstarts, wheatears and the Indian Robin. The chats derive their name from their ordinary note which is similar to the harsh chipping sound of two stones knocked together.

Blue-throat
Erithacus svecicus* or *Luscinia svecica
Nil-Kanthi or *Hussaini Pidda* in Urdu and Hindi
Dumbak in Sindhi
Nyul Hot in Kashmiri

Blue-throat, photograph by Mubashir Hasan

The Blue-throat is a widely distributed species through the greater part of Asia, northern Africa and Europe. It is found throughout Pakistan, Bangladesh and approximately in the western half of India. It migrates northwards in March-May towards its breeding grounds in Central Asia. The return migration begins in August and may last till October (1.8:219). It is a shy and skulking bird which feeds on ants, beetles, caterpillars and other insects. It is generally sighted moving with quick steps in a business-like manner on damp ground under bushes, seconds away from cover.

The Blue-throat is a sparrow-sized bird. The blue of its throat and breast has in its centre a small white or a rufous-and-white patch. On the belly side, the blue of the breast is bordered by three almost concentric half-collars, a narrow black border, a narrow white border and a broad rufous band. The belly is white and the upper plumage is brown. The female is like the male but the blue is replaced by buffish white and the rufous spot and the breast band are much reduced. In both sexes the black-tipped orange chestnut tail is conspicuous in flight.

Body length 144-158 mm.

Magpie Robin
Copsychus saularis
Dhaiyal, *Dayalchirya* or *Dhaiyar* in Urdu and Hindi

One of the most comely and dignified birds of the garden, the Magpie Robin is also one of the best of our songsters. Be it early in the morning or late in the afternoon, once the spirited, clear whistling tunes reach the human ear, it is impossible not to pause and try to look for the artiste, probably perched high up in a tree, a rooftop or some prominent place. In Pakistan it is found only in northern and central Punjab. It is widely distributed in India except in the far south and east.

The Magpie Robin is a bulbul-sized glossy dark blue-black and white bird frequenting groves and gardens. The lower plumage, a broad patch in the wings and the tail except the central feathers are white. The female is like the male but the blue-black is replaced by slaty above and grey on the throat and breast. It feeds on ants, moths, grasshoppers, snails, earthworms and small lizards; also other insects.

Body length 200-230 mm.

Magpie Robin, photograph by Mubashir Hasan

Black Redstart, photograph by Mubashir Hasan

Black Redstart
Phoenicurus ochruros
Thirthira or *Thirthira-kampa* in Urdu and Hindi
Thartharo in Gujrati

The Black Redstart breeds in mountainous areas: in western NWFP and Kashmir; northern Balochistan (between elevations of 2,100 and 3,100 metres), and the Himalayas (between 2,400 and 4,800 metres). The bird is widely distributed throughout Asia, Europe and parts of Africa. During winter it is found throughout Pakistan except the arid areas, and almost all over India, Bangladesh, Nepal and most of Burma. It feeds on insects, largely small beetles.

The Black Redstart is a sparrow-sized bird with a constantly quivering tail which has earned it the Urdu and Hindi name of the *Thirthira* (the quiverer). The male has a black throat and breast and the rest of its under parts are rufous. Above, the forehead and neck are black and the crown and lower back are grey. The rump and tail are rufous. The female is pale brown above and pale fulvous brown below. The tail is rufous.

Body length 145-155 mm.

Guldenstadt's Redstart
Phoenicurus erythrogaster

The Guldenstadt's Redstart is a very hardy species of high mountains and severe weather conditions. It breeds in Chitral, Gilgit, Baltistan, Ladakh and east through Nepal up to Arunachal Pradesh between elevations of 3,600 to 5,200 metres. During winter it migrates into valleys with elevations ranging from 1,500 to 4,800 metres (1.8:264). In summer it feeds on small beetles, ants and other insects and in winter mainly on berries (Roberts' personal comment).

The Guldenstadt's Redstart is a larger than sparrow-sized bird with a white forehead, crown and nape. The throat, breast, back and wings are black, the latter wear a large white path. The lower rump, tail and the rest of the under parts are chestnut. The female is pale brown above and pale fulvous brown below. The lower breast and flanks are ochraceous buff and the centre of the belly whitish (1.8:264).

Body length 170-190 mm.

Guldenstadt's Redstart, photograph by T.J. Roberts

Plumbeous Redstart
Rhyacornis fuliginosus
Kola Tiriv in Kashmiri

Plumbeous Redstart,
photograph by Mubashir Hasan

The Plumbeous Redstart is a sparrow-sized bird of the mountains. It breeds in the higher reaches of the river valleys in northern Pakistan from Chitral eastwards up to Kashmir at elevations of 1,200 metres (Chitral) and 4,400 metres (Manangbhot), from April to July (1.8:267). During winter it descends to the foothills. In India it is found along a narrow east-west band in the Himalayas and in Nepal. It is a winter straggler in north Bangladesh and India. It is also found in Vietnam and China.

It feeds mainly on insects found over or near mountain streams where it sits on a boulder or a protruding branch of a plant and makes sallies to snap up insects in the air or from the surface of water.

The female is dark grey-brown above. Below it is mottled slate and white. The male is entirely bluish slaty with a chestnut tail and a rufous belly. Young males often breed in female-like plumage (1.8:266).

Body length 120-140 mm.

Brown Rock Chat, photograph by Mubashir Hasan

Brown Rock Chat
Cercomela fusca
Shama or *Dauma* in Urdu and Hindi

The Brown Rock Chat is a sparrow-sized, plain brown bird with a darkish tail. At rest it could be confused with the female Robin, *Saxicoloides fulicata*, or with the female Blue Rock Thrush, *Monticola solitarius*. It is found in central Punjab, and is adapted to urban areas (2.2:130). In India, it is found in the Punjab, UP, MP, Bihar, Rajasthan and Gujrat (1.9:20). It frequents stony wastes, ravines, rocky hills, ruins, old buildings in villages, towns and cities (4:94). It feeds on beetles, ants and other insects.

Body length 165-170 mm.

Stone Chat or Collared Bush Chat
Saxicola torquata
Kharpidda in Urdu and Hindi
Dofa Tiriv in Kashmiri

The Stone Chat is very widely distributed in Asia, Africa and Europe. During winter, September to early April, it is found throughout Pakistan, Nepal, Bangladesh and almost the whole of India except its very southern tip. It is absent from western Balochistan and Sri Lanka. It breeds between elevations of 1,500 to 2,500 metres in northern Balochistan, Kohat, Chitral, Gilgit, Murree Hills and in the Himalayas from Ladakh to Bhutan.

The Stone Chat feeds mainly on ants, beetles and other insects. During winter it frequents damp areas, pasture land, fallow fields and tamarisk jungle. It is a dainty, sparrow-sized bird. The male has a black head, white collar, reddish breast and a white shoulder path. The female is rufous brown above, streaked with dark brown. The breast is rufous and below it is pale fulvous. It has a white wing patch.

Body length 125-130 mm.

Stone Chat, photograph by Mubashir Hasan

White-tailed Stone Chat, photograph by Mubashir Hasan

White-tailed Stone Chat or White-tailed Bush Chat
Saxicola leucura
Kharpidda in Urdu and Hindi
Dofa Tiriv in Kashmiri

The White-tailed Stone Chat looks like the Stone Chat, except that its tail is largely white and the rufous of the breast does not extend to the belly. It is confined all the year round to the proximity of the rivers of the Indus basin, as also of the lower Ganges and Brahmaputra systems. In all other respects it is the same bird as the Stone Chat, except for being sedentary and quite local in distribution.

Body length 125-130 mm.

Pied Bush Chat or Pied Stone Chat
Saxicola caprata
Kala Pidda in Urdu and Hindi
Pidda in Sindhi

The Pied Bush Chat is found from Iran to Indonesia. It is abundant in plains and lower hills in every type of country (4:85). It is a sparrow-sized, black-and-white bird; the male is jet black except for the belly, a wing patch and the undertail feathers. The female is dark brown with a rusty patch at the base of the tail. In the northern parts of the subcontinent where winter temperatures are low it is a summer visitor, arriving in February and March and leaving in September and October. In winter, it moves southwards right down to Sri Lanka.

It is common around cultivated fields, pasture land and scrub jungle and favours the vicinity of water. It keeps in pairs, generally the mates are never far apart. It is reported to be very territorial, actively resisting intrusion into its feeding or breeding area by its own species or other chats (1.9:33). It feeds on ants, beetles, larvae and some vegetable matter.

Body length 125-130 mm.

Pied Bush Chat, photograph by Mubashir Hasan

Grey Bush Chat or Dark Grey Bush Chat
Saxicola ferrea

The Dark Grey Bush Chat breeds between elevations of 1,500 and 3,300 metres throughout the forested regions of the Himalayas from Afghanistan and Chitral to Nepal, Bhutan and the State of Arunachal Pradesh in India. During winter, from September or October to March it moves down to lower altitudes of the foothills, all along the Himalayas. It feeds on insects and some seeds and is a familiar bird at the hill stations.

The male of the Dark Grey Bush Chat is pied black-and-white. The upper plumage is dark ashy-grey with black streaks; the wings are black edged with grey with a white patch. The entire lower plumage is white sullied with ashy. Of the female the entire plumage is rufous ashy, the tail and wings are brown, the sides of the head reddish-brown and the chin and throat are white.

Body length 150 mm.

Grey Bush Chat, photograph by Mubashir Hasan

Isabelline Wheatear
Oenanthe isabellina
Gidik in Brahui

Isabelline Wheatear,
photograph by Rolf Passburg

The Isabelline Chat is a sparrow-sized bird of steppes, deserts, stony plains and hillsides, feeding on beetles, insects and also on some seeds. Approximately from March to June, it breeds in the southern areas of the NWFP and considerably vast areas of adjoining Balochistan, Thal and Parachinar (NWFP) and in northern Balocistan. It also breeds in Turkey, through Central Asia up to Korea. In winter it spreads over all parts of Pakistan and north-western India. It has been seen east as far as Vanarasi and south as far as Pune. It has also been recorded in the Maldive Islands (1.840).

Its upper plumage is sandy brown, wings are dark brown with buffy edges, the tail is blackish brown while its base is white. The lower plumage is creamy buff. Both sexes look alike.

Body length 160-180 mm.

Desert Wheatear
Oenanthe deserti
Rann Piddo in Sindhi and Gujrati

The Desert Wheatear is found throughout Pakistan and the adjoining semi-arid areas of India from September/October to March/April. For breeding it migrates to Balochistan, the Karakorams, Tibet, Kashmir, central and south-west Asia and North Africa. It is a bird of arid plains and sandy wastes interspersed with cultivated fields or scrub land. Mostly it is seen on the ground or perched on a stone or a prominent place on a bush. It feeds mainly on insects and beetles.

The Desert Wheatear is an elegant, sparrow-sized, black and sandy coloured bird; in the male the throat, sides of the head and half of the tail are black, the wings are blackish-brown and the rest of the body is sandy buff, while the belly is buffish white.

Body length 150-170 mm.

Desert Wheatear, photograph by Mubashir Hasan

White-capped Redstart, photograph by Mubashir Hasan

White-capped Redstart or River Chat
Chaimarrornis leucocephalus
Gir-Chaondia in Urdu and Hindi
Chets Tal or *Kumidi* in Kashmiri

The River Chat is a dainty black and chestnut coloured, bulbul-sized bird with a white crown and nape. It is found along mountain streams of the Karakorams and the Himalayas from the NWFP in Pakistan through Kashmir and east up to the hills of Cachar and Manipur in India. It has also been reported from northern Balochistan (1.9:57) and from the Pakistan-Afghanistan border. It descends to lower altitudes during winter months.

The River Chat is an expert flier. It preys upon insects along torrential, often partially frozen streams, picking them up from fast flowing water as well as on the wing. Its wings, throat, sides of head and the tip of the tail are black; crown and nape are white and the rest of the body is rich chestnut. Both sexes look alike.

Body length 190 mm.

Brown-backed Robin or Brown-backed Indian Robin
Saxicoloides fulicata
Kalchiri in Urdu and Hindi

The Brown-backed Robin is sparrow-sized and is found in Pakistan, India, Sri Lanka and part of Bangladesh. It is not found in western NWFP and the northern regions of Balochistan. It shuns intensely cultivated irrigated tracts. It feeds on grasshoppers, ants, termites and other insects and their larvae.

The upper plumage of the male Brown-backed Robin is dark brown and the wings have a white patch. Its black tail keeps flipping up and down. The lower plumage is glossy blue-black and it is chestnut under the tail.

Body length 160 mm.

Brown-backed Robin, photograph by Mubashir Hasan

Blue-headed Rock Thrush, photograph by Syed Asad Ali

Blue-headed Rock Thrush
Monticola cinclorhyncha
Pala Tiriv in Kashmiri

A little larger than a bulbul, the Blue-headed Rock Thrush breeds in the Karakorams, the Himalayas and the Western Ghats and migrates to its winter quarters, central-west India, from September to April. Apart from its breeding areas in NWFP, Gilgit and Kashmir, it has been sighted in Pakistan during winter in Karachi and south of Lahore (1.9:68). It feeds on insects, occasionally on berries, seeds, flower nectar and also on animals such as frogs and lizards (1.9.68).

The crown, throat, shoulders and the back of the neck of the Blue-headed Rock Thrush are blue; the back is black and it wears a white wing patch; the rump and the rest of the under parts are chestnut. The female is plain olive-brown above and whitish and dark brown below.

Body length 170-180 mm.

Blue Rock Thrush, photograph by Tim Hurrell

Blue Rock Thrush
Monticola solitarius
Pala Tiriv in Kashmiri

The Blue Rock Thrush is a bulbul-sized bird. It is a bird of the rocks and if they are not available in the shape of hills and boulders, says Whistler, it finds a substitute in quarries, ruins of forts or unoccupied buildings (4:118). It is absent or very scarce in flat country. During summer it is found in its breeding areas in the mountain ranges of the Sulaimans, the Hindu Kush, the Karakorams and the Himalayas between elevations of 1,300 and 4,000 metres. It also breeds in the hills around Quetta and Chagai (Roberts' personal comments). During winter it is found in the foothills and further south right up to the Arabian Sea mainly along and west of the Indus. Roberts reports its occurrence in central and western Punjab, the NWFP and in the areas further north (2.2:164). It has been reported from Sindh, Bangladesh and much of India. The entire plumage of the male is dull dark blue, the wings and tail are brown. The female is grey-brown with dark streaks above and whitish cross-barred below.

The Blue Rock Thrush feeds on insects, occasionally on berries, seeds, and on animals such as frogs and lizards. It is silent in winter and in summer has a soft, melodious song, rather a whistle (1.9:75).

Body length 220-240 mm.

Black-throated Thrush or Dark-throated thrush
Turdus ruficollis atrogularis
Wanda Kastur in Kashmiri

The Black-throated Thrush is a myna-sized bird which is found in Pakistan in the riverain areas, western NWFP and north-eastern Balochistan from September/October to March. In India it is found along the Himalayan foothills and the adjoining plains and in Nepal, Bhutan and Bangladesh. During summer it migrates to Siberia for breeding (1.9:128). It feeds on

Black-throated Thrush, photograph by Mubashir Hasan

grasshoppers, beetles, caterpillars, ants, snails, earthworms, other insects, fruit and berries.

The male is grey-brown above, slightly spotted with a dark brown crown and back of the head; The throat and breast are black with whitish fringes and the rest of the underparts are white. The throat of the female is streaked with brown and whitish and the rest of the under parts are white. Above it is brown.

Body length 240-270 mm.

WARBLERS

Family
Sylviidae

The members of the *Sylviidae* family, commonly known as Old World warblers, number about 350 species in 63 genera. They are small or tiny—90 mm tall with fine narrowly pointed bills. Their plumage is generally dull and unobtrusive, greenish, brownish or greyish, barring some brightly coloured tropical species. Most warblers are solitary birds, many are arboreal and require for their habitat low bushes or reeds. Both sexes generally look alike. They feed mostly on insects, sometimes on berries.

On the grounds of broad similarities in ecology and habitat preferences, warblers may be classified as reed and bush warblers, tailor birds, leaf warblers, scrub and woodland warblers, grass warblers, tree warblers and others. Like many insect eating birds, most of the warblers that breed in the cold and temperate zones are migratory. Their southern destinations are spread far and wide in Africa, Asia and southern Europe. The Siberian Willow Warblers and European Arctic Warblers make journeys of up to 12,000 kms twice yearly to and from Africa and South-East Asia respectively (19:641).

The members of the *Sylviidae* family include some distinguished singers. Most species have well articulated songs which greatly help in distinguishing one race from another which otherwise are confusingly alike. Some are excellent mimics. However a minority actually warble, i.e., produce a continuous, gentle, trilling sound.

Streaked Wren-Warbler, photograph by T.J. Roberts

Streaked Wren-Warbler, Streaked Longtail Warbler or Graceful Warbler
Prinia gracilis
Khar Phutki in Urdu and Hindi
Door in Sindhi
Pitak in Brahui

The Streaked Wren-Warbler is found in northern Africa, south-west Asia, Afghanistan, Pakistan and northern India. It is a very small bird with pale brown upper plumage conspicuously streaked with dark brown. The lower plumage is pale fulvous. It has a long graduated, faintly cross-barred tail.

The Streaked Wren-Warbler prefers low sandy or uncultivated ground of bushes and reeds abundant in the riverain areas of the Indus and its tributaries as well as in the state land along the sides of the large irrigation canals. It is found in pairs or in small parties, not much above the ground, flying from one bush to another preying upon insects. All the time it flicks its wings, switches its tail from side to side with a slight upward flick (1.8:49). It feeds upon small beetles, caterpillars, grasshoppers and other insects. It breeds during March-April and in some areas in July-August.

Body length 100-130 mm.

Ashy Wren Warbler or Ashy Longtail Warbler
Prinia socialis
Phutki in Urdu and Hindi

The Ashy Wren Warbler is found in Pakistan only in northern and central Punjab. It is one of the commonest birds of India. It hops restlessly among grass stems and bushes with constant flicking of its erect tail. It feeds on larvae, spiders and insects, also flower nectar.

The Ashy Wren Warbler is a slightly less than sparrow-sized bird with a dark ashy-grey head, sides of neck and back, the rest of the upper parts are rufous brown. Its throat is whitish buff and the rest of the underparts are buff. Both sexes look alike. In winter, the tail is an inch longer.

Body length 120-130 mm.

Ashy Wren Warbler, photograph by Mubashir Hasan

Yellow-bellied Wren Warbler, photograph by F. J. Koning

Yellow-bellied Wren Warbler
Prinia flaviventris

The Yellow-bellied Wren Warbler is a sparrow-sized olive-green bird with a bright lemon-yellow belly. The forecrown and the sides of the head are dark grey. Both sexes look alike. It is a bird of the plains both of the central and southern Indus river system in the west and of the Brahmaputra and Meghna in the east. It is absent from Balochistan and most of India but has been reported from Burma and Vietnam. It feeds mainly on insects and frequents grasslands mixed with bushes in the vicinity of lakes, swamps, and backwaters of rivers. Its breeding season is mostly from June to October.

Body length 130-140 mm.

Tailor Bird
Orthotomus sutorius
Darzee, *Phutki*, *Piddi* in Urdu and Hindi

Like many other famous persons, says Whistler, the Tailor Bird is insignificant in appearance—a small green bird with a long pointed tail and a rufous crown (4:167). The entire lower plumage is shining white. For its nest, the bird first prepares an aerial cradle by sewing up two or more leaves together. The sewing is done with threads of cobweb, silk from cocoons, wool or cottons. Each stitch, which would pass well for human hands, is separate (4:168). The nest, a deep soft cup, made out of cottonwool and down, with a lining of a few horse hairs or grass stems, is placed in the cavity of the cradle. The normal clutch of eggs is three to four.

Barring deserts, thick forests and high hills and mountains, the Tailor Bird is one of the commonest birds of the subcontinent. However it is known more for its name than for its presence. It is an active, restless bird, hopping around bushes and creepers, investigating the flower-pots in the verandah, yet very few people take notice of it or care to recognize it. It is

Tailor Bird, photograph by Mubashir Hasan

absent from mountainous areas, most of Balochistan and south-western areas of the NWFP.

Body length 130 mm.

Blyth's Reed Warbler
Acrocephalus dumetorum
Podna or *Tiktiki* in Urdu and Hindi

According to Whistler, the neighbourhood of water has no special attraction for the Blyth's Reed Warbler; thus the name reed warbler can be confusing. All it requires is concealment in thick cover. There is a record of its breeding in northern Balochistan. In winter it is abundant in Pakistan, India, Bangladesh, Nepal and Sri Lanka. From March to May it is on its way towards its breeding grounds in Central Asia. The return migration starts in August and may last till October (1.8:110).

The Blyth's Reed Warbler is a sparrow-sized bird; the upper plumage is olive brown, the throat is white and the rest of the underparts are buffish. Both sexes look alike.

Body length 127.5-140 mm.

Blyth's Reed Warbler, photograph by Rolf Passburg

Desert Warbler, photograph by F.J. Koning

Desert Warbler
Sylvia nana
Phutki or *Pidda* in Urdu and Hindi

The Desert Warbler is a very small-sized bird, found throughout the valley of the Indus as well as in Balochistan, and Rajputana. It breeds in Central Asia and leaves for its breeding grounds in March and arrives back in September.

The Desert Warbler is pale greyish-brown above and creamy white below. Both sexes look alike. It feeds on insects and is usually seen singly running on the ground under the bushes or hopping among low scrub with the tail spread and partly cocked.

Body length 115-125 mm.

Lesser White-throat, Small White-throat
Sylvia curruca
Phutki or *Pidda* in Urdu and Hindi

The warblers under the name of White-throats are small-sized birds with white or creamish underparts and brownish or greyish upper parts. The head is a little darker and ashier than the back of the neck. Both sexes look alike. They are migratory birds, breeding in Central Asia while some breed in Siberia and Europe as well. They mainly feed on insects, and flit from bush to bush or hop on the ground.

The Lesser White-throat is found throughout Pakistan. It stays abroad approximately from April to October. It breeds in much of Asia and Europe except the far-northern and far-eastern areas of the former, also in parts of Balochistan and the NWFP (4:174). It is common in Sri Lanka and is patchily distributed in India. It is a very active, restless bird, continuously uttering loud tick-tick calls as it feeds (Roberts 2.2:245). It is dark greyish-brown above, the back and wings are tinged with brown. Below it is white, tinged with buff on the breast and belly.

Body length 130 mm.

Lesser White-throat, photograph by Mubashir Hasan

Western Crowned Leaf Warbler, photograph by Mubashir Hasan

Western Crowned Leaf Warbler or
Large Crowned Leaf Warbler
Phylloscopus occipitalis
Phutki in Urdu and Hindi
Viri Triv in Kashmiri

The Western Crowned Leaf Warbler, the commonest breeding bird throughout the northern forests of Pakistan, breeds in north-western Pakistan, Baltistan east through Kashmir from 1,800 to 3,200 metres (1.8:172), also in Afghanistan, Siberia and Central Asia. It starts migrating southwards and eastwards in August towards Bangladesh and central and south India where it is found during winter. Its return migration begins in March. Its double passage lies through the Punjab and northern India.

This warbler wears on its crown a pale median stripe between two darkish olive bands on either side; also it has two wing bars of which only one is sometimes visible. It is light greyish-olive above and whitish-tinged greyish on the breast and flanks, slightly streaked with yellow.

Body length 130 mm.

Greenish Warbler, photograph by Mubashir Hasan

Greenish Warbler, Greenish Leaf Warbler, Green Leaf Warbler or Greenish Willow-Wren
Phylloscopus trochiloides
Phutki in Urdu and Hindi

The Greenish Warbler breeds in north-western Pakistan, Gilgit east through the Murree Hills, Kashmir from 2,700 to 3,700 metres (1.8:167), also in Siberia and Central Asia. It starts migrating southwards and eastwards in August towards Bangladesh and central and south India where it is found during winter. Its return migration begins in March. On its passage it passes through all the provinces of Pakistan and northern India. It feeds on insects and frequents groves of large trees, gardens, orchards and to a lesser extent evergreen jungle.

The Greenish Warbler is dull green above and sullied yellowish-white below. It wears a dark streak through the eye, a pale yellow above the eye, and a pale yellow bar across the wings. Both sexes look alike.

Body length 110 mm.

Yellow-browed Warbler or Inornate Leaf Warbler
Phylloscopus inornatus
Phutki in Urdu and Hindi
Viri Triv in Kashmiri

The Yellow-browed Warbler is a very small-sized bird, two-thirds the body length of the sparrow. It breeds from May to July, throughout a large portion of Siberia and Central Asia (4:176); also in the Karakorams from the NWFP, Gilgit, Astore, parts of Hazara, Kashmir and Garhwal. For the rest of the year, it may be found in the plains of the subcontinent: in Pakistan in the northern areas of the Punjab and the NWFP and in the eastern, and central parts of southern India. It forages high up in the trees as well as in the bushes for insects.

The greenish colour, dirty white below, the double buffy white wing bars and the call note 'tiss-yip' are guides to its identity (4:176). It is called yellow-browed on account of the colour of the races which inhabit the Eastern Himalayas. It also has a broad buffy line over the eye. Both sexes look alike.

Body length 100 mm.

Yellow-browed Warbler, photograph by Mubashir Hasan

Chiffchaff, photograph by Mubashir Hasan

Chiffchaff
Phylloscopus collybita
Phutki or *Pidda* in Urdu and Hindi

The Chiffchaff is so called because of its song 'chiff-chaff-chiff-chaff, chiff-chaff'. Another of its notes is a very plaintive tweet. It is a very widely distributed bird, in a number of races, throughout Asia, Africa and Europe. In Pakistan it is found in Sindh, NWFP and the Punjab and in India in Rajputana and United Provinces from about September to the end of April. It breeds in Central Asia from the Urals in the west to the Kolymna River.

It is found singly or in parties of eight to ten, sometimes larger. It feeds on insects on the wing as well as on the ground. It has been seen to pick up insects from the water surface through a clever technique. The bird hops on to a loosely overhanging stem which slowly begins to go down due to the bird's weight. Having picked up the insect the bird hops off the stem without touching the surface of the water.

The Chiffchaff is pale olive brown above and a dull whitish below. Both sexes look alike.

Body length 110 mm.

—— FLYCATCHERS ——

Family
Muscicapidae

The members of the family *Muscicapidae* are small-sized birds with a broad flattened bill, suitable for feeding on spiders and insects, often caught on the wing, through constant sallies from prominent perches. Some of the species have well articulated songs.

Red-breasted Flycatcher or Red-throated Flycatcher
Muscicapa parva *or* **Ficedula parva**
Turra in Urdu and Hindi

Red-breasted Flycatcher, photograph by Brenda Wheeler

The Red-breasted Flycatcher is well distributed throughout Pakistan, India, Bangladesh and further east up to Vietnam and China. It is smaller than a sparrow and during March and May, migrates to Eastern Europe and Central Asia to breed. The return migration to the subcontinent starts in August and may end by October. It feeds on mosquitoes, midges and insects and is found in forest plantations, groves, orchards and clumps of large trees.

The chin and throat of the adult male is orange rufous, the rest of the underparts are white and the plumage above is pale brown. It usually carries its tail erect and wings partly drooping. The female has a whitish throat and a buffy breast. In their first winter the males are like females with a white throat and breast (Roberts' personal comments).

Body length 130 mm.

· FANTAIL FLYCATCHERS ·

Family Rhipiduridae

White-browed Fantail Flycatcher
Rhipidura aureola
Shamchiri, Nachan, Chakdil or *Machharya* in Urdu and Hindi

The White-browed Fantail Flycatcher is found all over the riverain areas of Sindh and the Punjab and eastwards up to Bangladesh. It is a bird of orchards, forests, groves, and well-watered cultivated areas with shrubbery and trees. It is a wonderful flier capable of making graceful loops, taking twists and turns preying upon insects on the wing. It frequents lower

White-browed Fantail Flycatcher, photograph by Mubashir Hasan

bushes but also descends to the ground often in the vicinity of grazing or resting cattle to snap up tiny insects on which it feeds.

It is a bulbul-sized cheery, restless, fantailed smoke-brown flycatcher with a broad white forehead and white under parts (1.7:207). Both sexes look alike.

Body length 170 mm.

▪ PARADISE FLYCATCHERS ▪

Family
Monarchidae

Paradise Flycatcher
Terpsiphone paradisi
Shah Bulbul, Sultan Bulbul or *Hussaini Bulbul* in
Urdu and Hindi
Fhambasir (Male), *Ranga Bulbul* (Female) or *Latraz*
in Kashmiri

Paradise Flycatcher,
photograph by Syed Asad Ali

Seeing for the first time a Paradise Flycatcher flying from the bough of one tree to another, in a Pakistani or an Indian garden, is an unforgettable experience. No other bird has the litheness and grace in flight that this flycatcher presents with its slender body followed by the long trailing streamers. As it chases its insect prey, the looping streamers present an exquisite spectacle. While perched it has a unique, breathtaking comeliness.

The Paradise Flycatcher observes a north-south migratory pattern. It may be found breeding throughout the valleys of the northern rivers from Chitral to Kashmir and eastwards up to Nepal; also in the foothill areas and the adjacent plains. Roberts has recorded breeding pairs as far away as Lahore and Khanewal. As a double passage migrant it is found all over the Punjab and Sindh and parts of Balochistan and the NWFP (2.2:314). In India it is common in the northern half of the Punjab and the Gangetic plains. It spends its winter in approximately the southern-half of the subcontinent (1.7:217). It is also found in Afghanistan and Turkestan. Its breeding season lasts from April to August.

Males more than four years of age have a conspicuous crest and their entire head and throat are black with a metallic blue sheen. Their wings are black-and-white and the rest of the plumage is silvery white. With streamers the length of the body may be as long as 500 mm. Up to the age of three years, the back, tail and streamers are rufous in colour. The females and first winter males have no streamers. They have an ashy broad collar on the throat and sides of the head.

The Paradise Flycatcher feeds on winged insects: flies, gnats, dragonflies, bugs, moths and beetles.

Body length 500 mm. with streamers.
Streamers length 300 mm.

BABBLERS

Family
Timalidae

The members of the family *Timalidae*, genus *Turdoides*, are terrestrial birds of pale brown plumage. They have a strong bill, legs and feet. Both sexes look alike. They always keep in small parties and are constantly on the move in the bushes, trees or on the ground with jerky movements and short flights. They feed on insects, fruit, flower nectar and small animals.

Striated Babbler
Turdoides earlei
Chilchil, *Bada-Phenga*, or *Genga* in Urdu and Hindi
Lelo in Sindhi

The Striated Babbler is found throughout the plains of the Indus river system but always in the proximity of water and swamps. It is also common in the plains of the Ganges and Brahmaputra along a line approximately parallel to the Himalayas covering the UP and Bihar up to the Burmese border.

This babbler is a bulbul-sized bird of reeds, high grass in swampy areas and bulrushes. It feeds on insects, snails and some vegetable matter, keeping in parties of six to ten. They keep moving, not staying at a place for long. As one bird flies away, the rest follow one by one or in twos. Jointly they defend their territory and protect members of the group against intruders. They are also known for keeping in body contact with each other while roosting at night.

Striated Babbler, photograph by Mubashir Hasan

The Striated Babbler is earth brown above with a streaked head and back. Below it is fulvous with dark streaks on the breast and throat. Both sexes look alike.

Body length 210 mm.

Jungle Babbler, Seven Sisters
Turdoides striatus
Satbhai, Ghoughai, Bhaina or *Pengya Myna* in Urdu and Hindi
Lailo or *Heddo* in Sindhi

The Jungle Babbler is a familiar, myna-sized bird of the Punjab, Sindh, parts of the NWFP, and much of the Indian Union. It is absent from elevations above 1,800 metres. Because of its habit of keeping in parties of six to twelve, it well deserves its Urdu name *Satbhai* (seven brothers), in English Seven Sisters. They chatter constantly and give alarm calls on noticing anything suspicious. All the members of the sisterhood roost sitting side by side with their bodies touching (1.6:226). It feeds on

Jungle Babbler, photograph by Mubashir Hasan

grasshoppers, ants, beetles and other insects. It is very fond of nectar and also eats grain and seeds.

The Jungle Babbler is grey-brown above and the lower plumage is paler, mixed fulvous and ashy. Its eyes are creamy white, bill and legs yellowish and the broad tail is tipped with white.

Body length 250 mm.

— TITS OR TITMICE —

Family
Paridae

The members of the family *Paridae* are small, active, short-billed birds of woodlands which often keep in flocks. A wide range is found throughout the Himalayas. Only seven different sub-species of *Parus* have been described as occurring in different parts of the subcontinent. Their diet is omnivorous, feeding largely on berries, nuts or any other fruit they can find usually fallen on the ground. In summers, their diet is largely insectivorous eating caterpillars, beetles, etc. Both sexes look alike.

Grey Tit or Great Tit
Parus major
Ranga Tsar or *Dantiwu* in Kashmiri

The Grey Tit is a sparrow-sized bird found in Afghanistan, northern and north-western areas of Pakistan; in central and southern India and along the Himalayan foothills; and in Bangladesh and parts of Burma. From October to March it migrates to its breeding grounds in Chitral and Ladakh. Its breeding has been reported by Roberts also from western NWFP and north-eastern Balochistan (2.2:361). During winter it spreads into the plains of northern Punjab. It is absent from Rajputana and the Gangetic plains (1.9:166). According to Whistler it is more a bird of the hills than of the plains. It is a cheerful bird and its loud whistle 'tsee, tsee, tsee' is always a cheery,

Grey Tit, photograph by Syed Asad Ali

welcoming sound. (4:20). It feeds on caterpillars and other insects, seeds, berries and flower buds and frequents edges of cultivation, bushes, gardens, lighter forested areas, and groves.

The upper parts of the Grey Tit are bluish ashy-grey. The tail is black and bluish ashy-grey. Its head, neck, breast and a broad line down the centre of the belly, are glossy black. There is a white patch on the cheek and a fainter one at the back of the neck; the remainder of the underparts are white-tinged with vinaceous.

Body length 140 mm.

SUNBIRDS

Family
Nectariniidae

The members of the family *Nectariniidae* are tiny arboreal birds, specially equipped with a long, slender, curved and finely serrated bill for feeding on flower nectar. They also have a long, tubular tongue adapted for sucking. The males are generally glistening purple, red, black, blue, green or yellow or a combination of these. They are expert fliers, capable of stationary flying or moving in the air in all manner of acrobatic positions and are thus able to extract the nectar even from the most inaccessible sites. They also feed on insects.

Purple Sunbird
Nectarinia asiatica
Shakar Khora, Phool Songhna in Urdu and Hindi
Dunbarg in Sindhi
Kala Pidda in Punjabi
Phul Chakli in Gujrati

The Purple Sunbird is found the year round, subject to local movements, all over Pakistan, India, Bangladesh and Sri Lanka. It is, however, absent from mountainous areas and from northern Balochistan and western parts of the NWFP. It feeds on the nectar of a variety of flowers and is very fond of fleshy blossoms of mhowa, *madhuka indica*, and of sugary exudation from Borassus palms (1.10: 36). Thus wherever there are flowers there is the Purple Sunbird.

Purple Sunbird, photograph by Mubashir Hasan

The Purple Sunbird is a restless bird, darting from branch to branch, flower to flower, industriously investigating, helping itself to whatever is offered, remaining on the move all the time. Satiated, it perches itself on a prominent place and sings. Espying an intruder, it readily darts away to drive it out of its territory.

Purple Sunbird in flight, photograph by Mubashir Hasan

In size it is smaller than a House Sparrow. The female of the Purple Sunbird is olive brown above and pale yellow below. The upper plumage of the male in its breeding season is dark blue and purple. The throat and breast are metallic purple, the belly dark purple and the sides are blue-green. The breeding season over, the upper plumage becomes pale olive brown, the wings and tail blackish and the lower plumage turns yellow with a broad blue-black band down the middle of the throat and breast.

Body length 100 mm.

── WHITE-EYES ──

Family Zosteropidae

Zosteropidae is the family of White-eyes or Silver-eyes; sparrow-sized birds known for a very distinctive ring of feathers around the eyes, emphasized by blackish loves and a narrow rim below the eye. A very widespread specie, it is found in Southern China, Formosa, the Phillipines and Northern Burma, the Indo-Chinese peninsula and over the whole subcontinent. Arboreal and gregarious except during the nesting season, they are generally encountered right up in the tree canopy and in small parties which keep up a contrast chorus of plaintive contact calls. They are omnivorous in diet, feeding upon seeds, berries, pollen, nectar and small insects.

White-eye
Zosterops palpebrosa
Baboona in Urdu and Hindi

In size a little smaller than the sparrow, the White-eye is a bird of trees and flowering shrubs. It frequents gardens, orchards and jungles close to cultivation for its food of insects, ants, seeds, small buds, nectar and berries. In Pakistan it is found in the Punjab and parts of Sindh and the NWFP. It is widely distributed throughout India, Bangladesh and Sri Lanka.

The White-eye keeps in pairs or in parties. It has a specially adapted bill and tongue for eating nectar and in the process acts as an agent in pollinating a wide variety of flowers (1.10:58). In

White-eye, photograph by Mubashir Hasan

its search for nectar, it dashes from flower to flower and from tree to tree and is often seen hanging upside down trying to reach an otherwise inaccessible flower.

The White-eye wears a prominent white eye-ring. The entire upper plumage and throat is greenish golden-yellow and the rest of the lower plumage is greyish-white. Both sexes look alike.

Body length 100 mm.

A closer view of the White-eye

ORIOLES

Family
Oriolidae

Oriolidae is the family of bright yellow, golden and black birds of myna size . They are known for their rich melodious calls as also for musical undertone soups. Most breed in Europe and are winter visitors to South Asia and parts of Africa. The head is uncrested and wings are long and pointed. The tail is much shorter than the wings and slightly graduated. Their diet is mainly insectivorous but they will take all kinds of fruit in the season.

Golden Oriole
Oriolus oriolus
Peelak in Urdu and Hindi
Poshinul or *Poshnul* in Kashmiri

The Golden Oriole is a beautiful myna-sized bird with a sweet melodious call. Very shy and secretive, it is a bird of gardens and forests and is more heard than seen as it keeps concealed in thickly foliaged trees. Also called the Mango-bird, it is purely arboreal, living on fruit, berries and flower nectar, but also takes insects. It breeds throughout Europe, Asia Minor, Central Asia, Afghanistan, Pakistan and northern India. According to Roberts, two races of the Oriole are found in Pakistan: one that winters in Africa and the other that migrates south-eastwards to India.

The male Golden Oriole is rich golden-yellow except for its wings and tail. The female and young male are greener yellow

Golden Oriole, photograph by Syed Asad Ali

with the underparts streaked with dull brown. It is a fast flier and bathes while flying by taking repeated dips at the surface of a pond and preening on a nearby perch (1.5:103). The male has a very pleasant and sweet loud call 'pee-ou-a', 'wiel-a-wo', 'pee-loo' and 'pee-loo-lo'.

Body length 250 mm.

SHRIKES

Family Laniidae

The members of the family *Laniidae* are myna to bulbul-sized land birds. Both the sexes look alike. Individuals have a fixed staked-out territory which is aggressively defended against intruders. Shrikes are good mimics of other birds' calls. They always perch on prominent lookouts from where they pounce upon large invertebrates and small vertebrates. When alarmed they use the trick of dropping from their perch, and flying off at high speed flat along the ground, to settle on another perch a hundred metres or so away. They are known to pirate morsels from other birds sometimes larger than themselves. What cannot be devoured forthwith is impaled on thorns or spikes for later consumption. The manner in which they hack and dismember their prey and store on thorns for future consumption what they cannot eat has earned them the English name of 'Butcher Birds'. In Urdu and Hindi, the Shrike goes by the general name of *Latora*.

Bay-backed Shrike
Lanius vittatus
Pachanak in Urdu and Hindi
Boro in Sindhi

The Bay-backed Shrike is a bulbul-sized bird found in open or cultivated country throughout the plains of the Indus valley. It can be regularly seen in parks and the greener areas of cities. It

Bay-backed Shrike, photograph by Mubashir Hasan

is not found in deserts or in northern parts of Balochistan. It is quite common in India.

The Bay-backed Shrike is a colourful bird. Its back is rich chestnut maroon, and the rump is whitish; below it is white, fulvous on the breast, pale rusty on the flanks and the head is white and grey.

General habits are as of other members of the family. Its breeding season lasts from April to July.

Body length 180-190 mm.

Rufous-backed Shrike
Lanius schach
Mattiya Latora or *Kajala Latora* in Urdu and Hindi
Hara Wataj in Kashmiri

The Rufous-backed Shrike is a large shrike, somewhat bigger than the myna. It is resident in open or cultivated country throughout Pakistan and India except in the arid areas.

Rufous-backed Shrike, photograph by Mubashir Hasan

General habits are as of other members of the family. The word *'wataj'* in its Kashmiri name means executioner, confirming the literal meaning of the English name 'Butcher Bird'. Its breeding season lasts from March to July.

It is rufous below; the chin, throat and upper breast are white; the crown, sides, hindneck and upper back are grey.

Body length 240-260 mm.

Great Grey Shrike
Lanius excubitor
Safed Latora, Dudia Latora or *Bara Latora* in Urdu and Hindi

Slightly larger than the Myna, the Great Grey Shrike prefers open semidesert country, edges of desert cultivation and thorn and dry deciduous forests. It feeds on ants, caterpillars, beetles, bugs and other insects as well as lizards, young mice and squirrels and young and sickly birds. In Pakistan it is found throughout the valley of the Indus but is only a summer breeding

Great Grey Shrike, photograph by Mubashir Hasan

visitor to Balochistan. It is found in central and southern parts of India.

The Grey Shrike is a black, grey and white bird with a longish black-and-white tail. Its wings are black with white patches; it is grey above and white below and it has a heavy hooked bill like that of a bird of prey.

General habits are as of other members of the family. Its breeding season lasts from March to June.

Body length 240-260 mm.

Shrike's Chicks

DRONGOS

Family
Dicruridae

Dicruridae is the family of drongos. They vary in size and are endowed with a glossy jet black colour; some species are slaty. They are arboreal and carnivorous. The tail is long and deeply forked with the male having a slightly larger tail and both species look alike. Feeding mainly on insects, capable of adroit aerial manoeuvres to seize flying insects as well as seizing much of its prey on or near the ground. Typically they can be seen perched on telegraph wires everywhere.

Black Drongo or King Crow
Dicrurus macrocercus or *Dicrurus adsimilis*
Kotwal, Bujanga, Buchanga, Kalkalachi or *Kalchiri* in
Urdu and Hindi
Japal Kalchit in Punjabi
Thampal in Pashtu
Kalaho or *Gohalo* in Brahui

The Black Drongo is one of the commonest birds of the countryside of Pakistan and India; is also found in Iran and Afghanistan. It has a graceful, sparrow-sized body and a deeply forked tail as long as the body itself. Its glistening black colour is glossed with blue. It is a brave and pugnacious bird having complete mastery over its flight.

In defence of their nest, a pair of drongos are occasionally seen driving away large birds like crows and raptors from their

Black Drongo,
photograph by Khan Mohammad

vicinity. Generally it is the crow which they attack from all sides, literally drawing circles around it, the crow not knowing which side to defend. 'Verily', says Whistler, the Drongo 'is the King of Crows'. It does not hesitate to attack the eagle, falcon or hawk with the same courage. Whistler's observation that a tree containing the Drongo's nest usually also contains the nest of an Oriole or a Turtle-Dove or some equally gentle bird, has led him to conclude that these species recognize the fact that the presence of the Drongo's nest above their heads is a guarantee of protection from all ordinary marauders. No wonder that in Urdu, the Drongo is also called *Kotwal*, that is the police officer incharge.

The Black Drongo is often seen riding cattle and pouncing upon insects disturbed by the latter's feet. But its favourite perch is a dead bough of a tree or a telegraph or electricity line from where it surveys and launches sallies to capture insects on the wing. Its food consists of dragonflies, crickets, grasshoppers, moths, bugs and their larvae.

The Drongo is one of the earliest birds to start calling in the morning and one of the last to fall silent in the evening. Its Punjabi names *kalcheet* and *kal-kali-chi* are a fair rendering of its calls.

Body length 310-330 mm. inclusive of the tail.

- CROWS, MAGPIES, JAYS -

Family
Corvidae

Members of the family *Corvidae* are the largest and most advanced perching birds. They are endowed with a high degree of intelligence and that has perhaps earned them simultaneously grudging respect and contempt from man. In Urdu and Hindi the crow is tauntingly termed a *Siana* (Clever-by-half). The Tree Pie is called *Lath* or *Mahalath* (Lord or Great Lord). In Sindhi, it is called *Malang* (Godly person who does not care about the world); and *Taka Chor* (rupee-thief) in Bengali. The White-rumped Magpie is called *Duzd* (thief) in Balochi. Indeed, the human response is a confused one.

They have a robust bill and legs and their food is practically all-embracing and they usually keep in parties. Both sexes look alike.

Black-throated Jay or Lanceolated Jay
Garrulus lanceolatus
Ban Sarrah in Pahari (Shimla)

The Black-throated Jay is a dove-sized bird affecting open mixed oak and conifer forests. It is found in the forested hills of the NWFP, along the northern half of the Pakistan-Afghanistan border, Murree Hills (2.2:418), Kashmir, and the hill ranges and valleys of the Himalayas between elevations of 1,500 metres to 2,500 metres, right up to central Nepal.

Black-throated Jay, photograph by Syed Asad Ali

The chin and throat of the Black-throated Jay are blue-black with broad white streaks; the wings are black and are closely barred with bright blue; the long blue-black tail is broadly tipped with white; the top and sides of the head are black and the overall body plumage is vinous grey. The head is crested and the throat feathers are long and pointed. Both sexes look alike. It is a trim and colourful looking bird on the whole.

Body length 225-235 mm.

Yellow-billed Blue Magpie
Cissa flavirostris (Ali and Ripley),
Urocissa flavirostris (Roberts)
Lotraza or *Literaz* in Kashmiri
Chainchal (Kangra)

The Yellow-billed Blue Magpie is a pigeon-sized bird with a long graduated tail of about 450 mm (18 inches). It is a bird of heavy jungle areas between elevations of 1,600 to 3,000 metres. It is found in the northern-most forested areas of Pakistan, in

Yellow-billed Blue Magpie, photograph by Syed Asad Ali

western NWFP and the Murree Hills (2.2:420). Confined largely to the Himalayas, it occurs right up to western Nepal. Its head, neck and breast are black; there is a white patch at the back of the neck; the entire upper plumage is purplish-blue and the lower plumage is white tinged with lilac.

It is a very active bird keeping in parties of seven to eight members. It feeds on fruits, nuts, insects, lizards, small mammals, eggs and chicks of small birds. It frequently feeds on the ground and while feeding holds the tail high, not letting it touch the ground, and adopts a hopping gait (4:11).

Body length inclusive of tail 630-650 mm.

Tree Pie or Rufous Tree Pie
Dendrocitta vagabunda or *Dendrocitta rufa*
Mahalath, Lath or *Mootri* in Urdu and Hindi
Mahtab, Mata, Malang or *Chand* in Sindhi
Khata Khan in Balochi
Aka Chor in Bengali

The Tree Pie is a bird of the plains and hill country of Pakistan, India, Bangladesh, and Nepal. It can be found up to the elevation of 5,000 feet wherever its food supply of fruits, nuts, insects, lizards, small mammals, eggs and chicks of small birds is available—generally in gardens and where large trees grow in clumps and avenues.

In Urdu and Hindi, the Tree Pie is called *Mahalath* (The Chief Lord, what the Governor-General or the Viceroy used to be called during the British Raj) or *Lath* (Lord or the Governor), presumably, because it is able to exercise almost sovereign authority over whatever a garden with fruit trees has to offer it. It has been seen tearing holes into nests of weaver birds and extracting and devouring the eggs and young. Mercilessly it devours not merely eggs and chicks of birds but also small sickly birds themselves. It eats snakes, wasps, bats—its diet being practically all-embracing (1.5:219). It also eats all kinds of fruit and those deputed to guard against pests are powerless against the Tree Pie.

The Tree Pie is a shy bird not having a very visible presence. Besides it can be excessively cunning and wary when occasion

Tree Pie, photograph by T.J. Roberts

dictates (1.5:219). It has a repertoire of harsh and raucous as well as melodious calls.

The head, neck and breast of the Tree Pie are brownish-black; the wings are dark brown with a large greyish-white patch on the sides; the belly is bright rufous and the tail is graduated greyish with broad black tips. Both sexes are alike.

Body length 500 mm. including the tail 300 mm.

House Crow
Corvus splendens
Kowwa, *Desi Kowwa* or *Kag* in Urdu and Hindi
Kan in Sindhi
Kav in Kashmiri

One of the most common and familiar birds in the land, the Crow is unanimously considered by Urdu and Hindi speaking people as the most *siana* among the birds, which might mean one or more of the following: highly alert, wisely inquisitive, deliberately impudent, consciously clannish with a capacity to distinguish between a harmless human from the one not to be trusted. Besides intelligence and a limited capacity for reasoned thought, Ali and Ripley contend that the crow possesses a distinct sense of humour (1.5:245). They cite as evidence some antics of the crow, such as tweaking tails of other birds or ears of sleeping cows or dogs (1.5:245) or hopping on or off the backs of well gorged vultures sitting on the ground (4:7), with no apparent object other than to enjoy their annoyance and discomfiture. There is one living creature, however, which exists on account of its ability to deceive the crow. It is the Koel, which successfully either lays or substitutes its eggs in the nest of the crow and thus makes it rear its young.

The crow eats everything that man would eat or leave uneaten and more. It is bold and thievish, and robs from unattended shops in the bazaar or snatches sweetmeats off the trays of the vendors, and is a first rate municipal scavenger. The Crow is

House Crow, photograph by Mubashir Hasan

said to have a very large and eloquent vocabulary to express finding of food, alarm, anger, suspicion, contentment or distress. 'Yet with all their villainies', says Whistler, 'there is much that is attractive about the sleek, intelligent, shameless bird that is the companion of our daily life in India.'

The crow is an expert flier capable of somersaulting, carrying out lightning twists and remarkable acrobatics. The crow often commutes long distances from its communal roosts. Vast numbers, running into thousands, leave their roost of clumps of large trees or forest plantations early in the morning in all directions to return at sunset.

The crow is monogamous. Mr and Mrs Crow, even in non-breeding season, will often sit, lovingly snuggled together, on a shady branch during the day and scratch each other's lowered head. They evidently pair for life.

The crow is a highly glossed black bird, the gloss having a purplish sheen. The collar, breast and upper back are grey. It is found up to the elevation of 1,200 metres throughout Pakistan, India, Bangladesh and up to Vietnam. Its several races are distinguished from one another by slight variations in the shade of the non-black portions of the plumage (4:5). Both sexes are alike.

Body length 430-450 mm.

Jungle Crow, Long-billed Crow or Himalayan Jungle Crow
Corvus macrorhynchos
Pahari Kowwa or *Jangli Kowwa* in Urdu and Hindi
Kav or *Diva Kav* in Kashmiri

The Jungle Crow is a bird of mountains and forests. In size it is larger than the House Crow and is found from Balochistan north through Afghanistan and the NWFP and eastward along Chitral, Gilgit and Kashmir and the Himalayan region of India, Nepal, Sikkim and Bhutan, between elevations of 1,800 to 4,500 metres

(1.5:252). It is a uniformly black bird with a metallic purplish sheen and a heavy black bill.

It feeds on birds' eggs, young and sickly birds, chicks of poultry, rats, mice, lizards, carrion, garbage and offal, frogs, insects, wild and orchard fruits, cereals—practically everything animal as well as vegetable.

It is curious and inquisitive but less sophisticated and cunning than the House Crow. Whistler saw it settling on packs of mule trains crossing the high passes, travelling with them and tearing holes in the packs to get at the contained corn (4:4). They roost communally along with House Crows, mynas and other birds. It is an expert flier and indulges in acrobatics and is fond of soaring and circling at great heights. Its call is more raucous and guttural than the House Crow's.

Body length 430-500 mm.

Jungle Crow, photograph by Mubashir Hasan

— MYNAS, STARLINGS —

Family
Sturnidae

The members of the *Sturnidae* family are medium-sized land birds with a short tail, strong legs and bill. The family copmrises about 110 species, in 25 genera (19.562). On familiar terms with man in towns as well as the countryside, some of the species are among the most common of the birds and are found throughout Asia, Europe and much of Africa. They are mainly arboreal species frequenting forest edges while others inhabit grasslands. Most eat fruit and insects while starlings are omnivorous. Generally found in pairs or in larger numbers, they walk and run on the ground but do not hop. The starlings found in Pakistan are migratory while the mynas are resident.

Brahminy Myna, Brahminy Starling or
Black-headed Myna
Sturnus pagodarum
Brahminy Myna, *Kalasir Myna* or *Popoya Myna* in
Urdu and Hindi

The Brahminy Myna is a bird of open cultivated areas endowed with shrubbery and trees. It avoids arid, semi-desert and desert tracts. In Pakistan it is found in the valleys of the northern rivers in the NWFP and Kashmir, mainly between 900 and 1,800 metres. It has also been reported from Sialkot district. It is common in India. It has a glossy black forehead; the upper plumage is grey and below it is rich buff. Both sexes look alike.

Brahminy Myna, photograph by Syed Asad Ali

It feeds on fruits, berries, nectar of flowers, and freely associates with other mynas to prey upon grasshoppers, moths, caterpillars and other insects. It roosts along with other mynas and birds in large leafy trees in extremely noisy gatherings. It is a good songster and is also a good mimic, capable of learning tunes and calls of other birds.

Body length 220 mm.

Common Starling, Sindh Starling or Starling
Sturnus vulgaris
Kala Tilyar, Tilyar Myna, Tilora, Nakshi Tilyar in
Urdu and Hindi
Karo Whahio in Sindhi
Tsininhangoor in Kashmiri

The Common Starling is so called as it is the commonest amongst its several races. It feeds on fruits, berries and insects and is found in cultivated areas, in the vicinity of habitations and in grazing lands around lakes. During May/June, they breed in Europe, Asia Minor and much of Central Asia. The annual migration from the subcontinent takes place through Chitral and Gilgit in mid-March to mid-April and then again in October/ November. The birds from Pakistan predominantly are migrants breeding in Siberia. Roberts also reports six breeding areas in Pakistan for the very local Sindh sub-species (2.2:450). It is common in India.

The entire plumage of the Common Starling in the breeding season is black, iridescent with a high gloss of red, purple, green and blue. Following breeding, the feathers of both sexes as winter advances abrade and the buff spots disappear (Roberts' personal comments).

Like others of its tribe, the Common Starling is found in flocks, small or large. According to Whistler, the chief characteristic of the flocks is hurry. Feeding on the ground, digging into the crevices of the soil and extracting the various harmful grubs and insects on which they feed; all the time, the flock advances with a bustle and hurry (4:199). A feeding flock will, for no apparent reason discernible to man, often take to the

Common Starling, photograph by Mubashir Hasan

air and fly into a nearby tree for a brief interval, before gliding
back in twos and threes to the same feeding area.

Body length 190 mm.

Rosy Starling or Rosy Pastor
Sturnus roseus
Gulabi Tilyar, *Gulabi Myna* or *Tilyar* in Urdu and Hindi
Waheeo, *Wyho* or *Wyha* in Sindhi

The Rosy Starling is most visible all over the valley of the
Indus and Balochistan during its spring and autumn migrations
to and from its major winter quarters in India where it is
common. It is a rare straggler in Bangladesh. They are, however,
only totally absent from Pakistan between late May and early
July and may be sighted from August to March, even July to
April (1.5:164). Roberts comments on the interesting
phenomenon that this species spends nearly ten months inside
Pakistan or India, yet breeds as far as 4,000-5,000 kilometres
away. In 1926, a nestling ringed in Hungary was recovered in

Rosy Starling, photograph by Mubashir Hasan

Lahore at a distance of 4,800 km. However, evidence of breeding has come from north-eastern Afghanistan, from as close as 100 km from Pakistan's border (2.2:463). For years, on the ninth day of April, the flocks of the Rosy Starling have not disappointed this author. There they are, on the majestic avenues of the pipal trees, *F. religiosa* on the Mall Road of Lahore. Sadly the number has been dwindling from year to year.

The Rosy Starling is famous as a great chatterer. It always keeps in flocks, often large, sometimes of swarm proportions. The unceasing and deafening din and clamour of its chatter on a large fruiting tree or an avenue is totally overwhelming.

The Rosy Starling is a two-colour bird with a glistening black head, neck and upper breast, wings and tail, the rest of the body being rose pink. Both sexes look alike. It feeds on fruits and berries, flower-nectar and insects and often preys upon insects disturbed by the feet of grazing cattle. The fruiting pipal and flowering Flame-of-the-Forest trees are a must for its visit. It is highly beneficial as a wholesale destroyer of locusts and as an important agent of seed dispersal.

Body length 230 mm.

Pied Myna or Asian Pied Starling
Sturnus contra
Ablak or *Ablaki Myna* in Urdu and Hindi

It is only recently that the Pied Myna's sightings are being reported from Pakistan. Ali and Ripley mention (1.5:173) its occurrence only to the south and south-west of Ludhiana (Indian Punjab) and Hissar (Haryana).

It keeps in small flocks and is rarely met away from human habitations. It feeds on insects, fruit and grains of cereal.

The Pied Myna is an elegant black or blackish-brown bird with a white belly, a broad white bar on the wings and a white patch through the eye backwards.

Body length 230 mm.

Pied Myna, photograph by Mubashir Hasan

Myna or Common Myna
Acridotheres tristis
Myna or *Desi Myna* in Urdu and Hindi

The Myna is found throughout South and South-East Asia. An established part of folklore, it is one of the most common and well-known birds of Pakistan, India and Bangladesh, but is absent from most of Balochistan and parts of the NWFP.

A child with a sweet way of talking may be lovingly called a myna among Urdu and Hindi speaking people. It is an immaculately groomed bird. In the olden days, when a raised platform or a bench served as a perch to wash the face of a child with a *lota* or *garwi* of water, the brat, not cooperating with the mother, was told to take a look at the Myna, invariably present in the inner courtyard of the house, as an example of a finely washed and dried face and immacutely dressed hair.

The Myna is a tame and confiding bird, fearlessly walks the lawns, enters verandahs and yards to pick up grain or crumbs. It feeds on grain, fruit, insects, mice, lizards, and almost all items of human food, and also visits refuse dumps to pick up tidbits discarded by man. In cultivated areas they follow the plough or moving cattle to prey upon the insects disturbed by the movement.

The Myna has a dark brown body with a glossy black head. It has a large white patch on the wing and its legs and bill are bright yellow. Both sexes look alike. It is a sociable bird, generally keeps in parties of five or six. Often they gather into large flocks, especially in the non-breeding season, and roost along with other mynas and birds in large leafy trees in extremely noisy gatherings before finally settling down to sleep. The silence is often broken for short durations even during the night for no discernible reason.

When not in a hurry, the Myna has a slow and measured gait, termed as the one with *matak* in Urdu, each step involving a deliberate, stylish movement of the neck along with a mild swagger of the body. It gives wild alarm calls at the sight of a cat or a snake which is heeded by birds in the vicinity and is worth investigating by man. In captivity it is capable of repeating a few words of human speech.

Body length 230 mm.

Myna, photograph by Mubashir Hasan

Bank Myna
Acridotheres ginginianus
Lal Chonch Myna or *Ganga Myna* in Urdu and Hindi
Lali in Sindhi

The Bank Myna is found throughout the plains of the Indus but is most typically associated with the main rice-growing tracts (Roberts' personal comments); it is absent from most of Balochistan, parts of the NWFP and the northern area of Pakistan (2.2:468). It is common in northern India and Bangladesh. It keeps in flocks even during the breeding season (1.5:181) and freely associates with other mynas and birds in looking for food and in roosting. Compared to the Common Myna it is more a bird of open cultivation and the countryside (4:205). Its habits are not markedly different from the Common Myna.

The Bank Myna has a slaty-grey or bluish-grey body, the head and neck are black, and the naked skin below and behind the eye is brick red; and so is the base of the bill instead of yellow in the Common Myna. Both sexes look alike.

It feeds upon fruit, grain and insects. It nests colonially, almost exclusively in earthen banks and cliffs in holes which it excavates for itself, always in the vicinity of water (4:205). The tunnel leading to the nest may be one-half to one metre deep, slightly upward sloping.

Body length 210 mm.

Bank Myna, photograph by Khan Mohammad

SPARROWS

Family
Passeridae

The members of the family, *Ploceidae* and *Passeridae* comprise sparrows, rock sparrows, snow finches, bayas, and weaver birds. Not long ago the members of the family *Passeridae* were considered to be members of the sub-family *Passerinae* of the family *Ploceidae*. Thinking among scientists seems to have changed. *Passerinae* has been upgraded into a full-fledged family in its own right with sparrows and snowfinches as its members. The House Sparrow is perhaps the most cosmopolitan and possibly the most numerous bird in the world.

The birds of these two families are equipped with a thick bill, short and stout legs and powerful wings and they are mainly vegetarian feeding on grains, grass seeds, berries and insects. Their plumage varies from grey and brown to various shades of red, yellow, green and white.

House Sparrow
Passer domesticus
Chirya, Gharello Chirya, Goriyya, Churi or *Khas Churi* in Urdu and Hindi
Chiri in Punjabi
Ginjishki in Balochi

Originally a bird of Asia and Europe, the House Sparrow has been introduced in all the continents except Antarctica. It is abundant in all the countries of the subcontinent. Man and

House Sparrow, photograph by Mubashir Hasan

sparrow are great eaters of cereal grains. The sparrow's food supply is generally connected in some way with man. The larger and more prosperous a city or village the more the sparrow flourishes (4:227). The Sparrow is well established in the folklore of the peoples of Asia. In Urdu and Hindi, as surely as in other languages of the subcontinent, there are a number of stories for children and proverbs for adults concerning the sparrow.

Long ago, Salim Ali wrote that for India as a whole, the House Crow and the Sparrow would be hard to beat for commonness and abundance (*The Book of Indian Birds*, 1972, p.xxi). For the agricultural areas of the Indus valley, the House Sparrow is likely to leave the Crow far behind in any population census of the birds. Roberts (1.1:43) reports of a study, made in 1980, of a small village in Rahim Yar Khan which revealed that during March and April, its human population was 416 persons and that of the resident sparrows was 16,700. For every household there were 321 sparrows.

When foraging a ripened crop, sparrows can collect in very large flocks; usually it keeps in parties except during the breeding period when it keeps in pairs raising several successive broods (1.10:66). Random but close observations made by this author, while detained in a Lahore jail by the dictator General Ziaul Haq, revealed the following: (1) On two occasions it was found that it was the male who first selected the site of a new nest and initiated the nest building process. The female took her time to approve of the site and join the process of nest building. The first straw she brought was transferred to the bill of the male, who took it inside the hole. (2) The male of a pair engaged in raising a brood hit a ceiling fan and died. Within a period of three days, the female who was still feeding the youngest of the chicks, was apparently joined by a new mate who selected a new site, a short distance away from the old one, and the female joined in building a new nest. (Mubashir Hasan, 1978, *Razm-e-Zindigi*, Nafees Printers, Lahore).

The top of the head of the male sparrow is ashy-grey; the back and shoulders are chestnut streaked with black; the wings have bars of chestnut and dark brown; there is a black patch from the chin to the upper breast and the cheeks and the remainder of the lower plumage is white. The upper plumage of the female is greyish-brown streaked with fulvous and dark brown on the back, the lower plumage is brownish-white.

Body length 150 mm.

Spanish Sparrow or Willow Sparrow
Passer hispaniolensis

The Spanish Sparrow is widespread in the NWFP and Punjab from September to April. It is only a passage migrant in Balochistan. Its flocks also invade the Punjab, Haryana, and the adjacent areas of the UP in India. It breeds in Afghanistan, Iran, Central Asia, Turkey, North Africa and Spain.

As the wheat crop ripens in the plains of the Indus valley, very very large foraging flocks are encountered, especially near

Spanish Sparrow, photograph by Syed Asad Ali

their roosting sites among the reeds in the riverain areas. They feed on grains, seeds and insects.

The male of the Spanish Sparrow has a chestnut crown and back of the head; the wings are pale brown with chestnut and whitish bars; the cheeks are white; the throat and breast are black and the rest of the lower plumage is whitish, streaked with black on the flanks. The female is not much distinguishable from the female of the House Sparrow. It is greyish-brown, streaked with fulvous and dark brown on the back; its lower plumage is plain brownish-white.

Body length 150 mm.

Sindh Jungle Sparrow
Passer pyrrhonotus

The Sindh Jungle Sparrow is a little smaller in size than the House Sparrow. Barring the northern districts of the Punjab, it is found throughout the plains of the Indus Valley, inclusive of the

Sindh Jungle Sparrow (below) with white-cheeked Bulbul (above), photograph by Mubashir Hasan

plains of the Indian Punjab. Essentially, it is a riverain sparrow frequenting tall grass and bushes along the rivers and lakes and in semi-desert localities but in the vicinity of water (1.10:73). It feeds mainly on grass and weed seeds. Its breeding season lasts from the end of February to September.

The male of the Sindh Jungle Sparrow is similar to that of the House Sparrow except that its breast is not black, only the chin and throat are. Its lower plumage is pale ashy. The female is not distinguishable from that of the House Sparrow.

Body length 120 mm.

Cinnamon Tree Sparrow
Passer rutilans
Lal Gouriyya or *Lal Chirya* in Urdu and Hindi

The Cinnamon Tree Sparrow is found from Chitral to Nepal in the Karakorams and the Himalayas between elevations of 1,200 and 2,400 metres (1.10:75). It breeds in Chitral (Kafiristan), Tibet, China, Japan and eastern Afghanistan. It is a forest sparrow (4:228) which feeds mainly on grains, seeds, insects and berries. During winter it descends to lower altitudes, sometimes in the foothills and the plains.

The upper plumage of the male of this Sparrow is bright rufous chestnut streaked with black on the back; the chin and

Cinnamon Tree Sparrow, photograph by Mubashir Hasan

the centre of the throat are black and the rest of the lower plumage is greyish-yellow growing yellower towards the tail. The upper plumage of the female is brown, streaked with dark brown on the back. The lower plumage is pale ashy-yellow.

Body length 150 mm.

Yellow-throated Sparrow
Petronia xanthocollis
Raji or *Jangli Chirya* in Urdu and Hindi

The Yellow-throated Sparrow is found throughout the riverain plains of the Indus, Rajputana and central and southern India. It is also found in Iran, Afghanistan, and Iraq. It is a bird of dry deciduous forests and areas scattered with thorn jungle. It feeds upon seeds of grass and weeds, vegetable matter, berries, nectar, grains, ants, beetles, caterpillars and other insects. It nests in trees. The breeding season lasts from the end of February to May (1.10:81).

Yellow-throated Sparrow, photograph by Syed Asad Ali

The lower plumage of the Yellow-throated Sparrow is pale grey-brown with a yellow patch on the throat; the chin and belly are whitish. The upper plumage is grey-brown with a chestnut shoulder patch and two whitish wing bars. Both sexes look alike.

Body length 140 mm.

WEAVER BIRDS

Family Ploceidae

Baya
Ploceus philippinus
Baya in Urdu and Hindi

The Baya is famous not merely for the astounding skill and dexterity it shows in making its nest but for being a lovable pet capable of learning to perform. Throw a ring in a well, the Baya will go after it like a bullet and will retrieve it before it is lost in the water. It can thread tiny beads with a needle or pluck and bring back for its trainer, leaves from a chosen tree. It obviously

Baya, photograph by Syed Asad Ali

shows skill as well as intelligence (1.10:92). The male Baya in breeding plumage is called *Chand Baya* in Urdu and Hindi.

In Pakistan, the Baya is found all over the Punjab and Sindh but is absent from Balochistan and most of the NWFP. It is found throughout India, generally up to the elevation of 1,200 metres. The Baya frequents open cultivated areas, grasslands and scrub with babool, date or palm trees in the vicinity. It feeds mainly on grass and weed seeds, cereal grains, insects and flower nectar.

Nests of the Baya, grouped in colonies and suspended from babool trees along the banks of canals, country roads, or in clumps of trees adjoining scrub or cultivated fields are a common sight in the plains of the Indus. The nest, a marvel of ingenious construction never ceases to amaze anyone who cares to examine it. The long sleeve-like entrance tube leads to the egg chamber. The entire structure is suspended from a fragile-looking stem, practically unapproachable by a predator. This author saw in 1981, the remains of a cat which had tried to approach a baya's nest hanging at the end of a branch of an acacia tree. Apparently, the cat having reached within about 300 mm of the nest, had gotten to a point from where it could proceed neither forward nor backward and died of starvation, if not of cardiac arrest.

The breeding biology of the Baya requires of the male of the species not only to build the structure of the nest but also to 'sell' it to a fastidious female. As the adult male, donning its full breeding plumage, proceeds with nest building, while other males are doing the same on other twigs of the same tree or in its vicinity, the prospective wives begin visiting what is ultimately to become a colony. They inspect and examine the quality and adequacy of the construction and listen to the song and watch the pleading gestures of the males. When a female approves of and selects a nest, it chooses a husband. The male completes the nest, instals the female on a clutch of eggs and proceeds to build another nest close by for yet another female. It may end up establishing two, three, even four households in a season. The female incubates the eggs and raises the chicks, though the male may occasionally help in feeding.

The upper plumage of the male in the breeding period is dark brown streaked with yellow on the back; the crown and breast are also yellow. The rest of the lower plumage is cream buff. The crown and back of the female are yellowish buff streaked with dark brown. The throat is white tinged with yellow; the breast is yellowish buff with brown streaks on the sides and the belly is cream buff. The male is similar to the female in the non-breeding period, but darker and more boldly streaked and its breast is very pale yellow (1.10:89).

Body length 150 mm.

Streaked Weaver Bird
Ploceus manyar
Telia Baya in Urdu and Hindi

The Streaked Weaver Bird lives and nests in reed beds. It is more of a water haunting species feeding on seeds or less

The Streaked Weaver Bird busy completing his nest, photograph by Mubashir Hasan

commonly, insects found in the grass (4:212). It also eats cereal grains. In Pakistan it is found in the Punjab and Sindh. It is widely distributed throughout India except for the desert areas of Rajasthan.

Like the Baya it keeps in large flocks in and out of the breeding season. It also breeds in colonies. The nest is not as elaborate as that of the Baya. The male takes some part in the incubation duties. Both the sexes take care of the young.

The upper plumage of the male is dark brown streaked with fulvous and the sides of the head are dark brown; the crown is golden-yellow. The throat is brown and the lower plumage is buff streaked with dark brown, heavily on the breast.

Body length 150 mm.

MUNIAS

Family Estrildidae

Mumas, waxbills, grassfinches and mannikins are birds of the *Estrildidae* family. The are found all over the subcontinent extending eastwards to the Malaysian and Indo-Chinese sub-mountain regions. They are brightly coloured birds, rather tame and confiding. They are very popular as cage birds and are heavily persecuted by professional bird catchers for the aviary trade, both local and for export to the Middle East and Europe. They are granivorous in diet, foraging mainly on grass, reed seeds, etc. and learn to eat bread crumbs and cereals easily. Some are also known to take small ants and beetles.

Red Munia, Red Avadavat or Avadavat
Estrilda amandava* or *Amandava amandava
Lal or *Lal Munia* in Urdu and Hindi

Smaller in size than the sparrow, the Red Munia is a red jewel of a bird much admired for its bright and lively nature. It feeds mainly on grass seeds and frequents reeds and tall grass near marshes, lakes or swampy ground. Except for the drier and mountainous regions, the Red Munia is found throughout Pakistan, the Indian Peninsula and most of Bangladesh. It keeps in pairs or small flocks and roosts communally in reed beds or sugarcane fields (1.10:102).

Red Munia, photograph by Mubashir Hasan

During the breeding period, the upper plumage of the male of the Red Munia is brown to crimson, the wings are dark brown with a few white spots, and the tail is dark brown. The lower plumage is crimson spotted with white, while the belly is dark brown. In the non-breeding period the male is like the female but its throat and breast are greyer. The upper plumage of the female is brown with a few white spots on the wings. The throat and breast are greyish buff and the lower belly is buff.

The Red Munia is a persecuted bird in Pakistan. It is caught in the tens of thousands annually by professional netters for export or to be sold locally as a cage-bird. Roberts has noted that this trade has reduced the population of the species and it has now become quite scarce.

Body length 100 mm.

White-throated Munia
Eodice malabarica or *Lonchura malabarica*
Churchura, Charakka, Charga or *Pidda* in Urdu and Hindi
Pavai Munia in Gujrati

Smaller in size than the sparrow, the White-throated Munia feeds mainly on grass seeds, ants, beetles and other small insects. For lack of migration and for wearing the same unchanged plumage throughout the year, the White-throated Munia appeared to Whistler to be a dull bird (4:213). It prefers drier country than other munias, frequents babool jungle, grassland, and sparsely scrubbed land. Barring parts of Balochistan and the NWFP, the White-throated Munia is found throughout Pakistan and throughout central and south India. It is also found in the southern Arabian Peninsula. It keeps in flocks and roosts in parties of five or six huddled together in old nests (1.10:105).

The upper plumage of the White-throated Munia is brown, the wings and tail are dark brown; the lower plumage is creamy white, flanks brownish buff and the rump is white. Both sexes look alike.

Body length 100 mm.

White-throated Munia, photograph by Syed Asad Ali

Spotted Munia
Lonchura punctulata
Telia Munia or *Seenabaz* in Urdu and Hindi

The Spotted Munia is most numerous in open country where scrub jungle alternates with cultivation and the vegetation is luxuriant (4:215). It feeds on grass seeds, berries and rice. In Pakistan it is found along a narrow south-eastern north-western band with the Lahore-Kasur area at one end and some point in Swat, NWFP at the other (2.2:507). The photograph shown in this volume was taken in Mr Raza Kazim's bungalow in Gulberg, Lahore, where a flock of about twenty individuals stayed for several days. In India it is found in the Gangetic plains and south and south-westwards up to the end of the Peninsula. It is common in Sri Lanka and Bangladesh.

The spotted Munia is smaller in size than the Sparrow. The sides of its head, chin and throat are rich chestnut. The belly is white, each feather edged with black giving a scale-like pattern. The upper plumage and wings are light chocolate. Both sexes look like.

Body length 100 mm.

Spotted Munia, photograph by Mubashir Hasan

FINCHES

Family Fringillinae

The members of the family *Fringillinae* are seed-eating, sparrow-sized birds associated with trees. They have stout conical bills, strong skulls, large jaw muscles, powerful gizzards and they specialize in eating hard seeds by first crushing, then peeling the husk, before swallowing the kernel. Although their main food is seeds, the *Fringillinae* feed the young entirely on insects, specially caterpillars, while the *Carduelinae* feed their young either on a mixture of seeds and insects or on seeds alone. Many species of finches have special song flights over their breeding areas (19:215). They have a characteristic bounding or dancing flight. The finches are very familiar birds in Europe. They come in all colours, in combinations of red, grey, green, blue, black, purple, yellow, brown or white. For centuries, finches have been kept in cages as pet birds on account of their bright colours, engaging songs and for the ability of some species to breed in captivity.

Red-frontcd Scrin or Gold-fronted Finch
Serinus pusillus
Tiok in Kashmiri

The Red-fronted Serin breeds in mountain ranges of the Sulaiman, the Karakorams and the Himalayas up to Garhwal. During summer, it breeds in northern Balochistan, Chitral, Kaghan, Gilgit, Baltistan, Kashmir, Ladakh, Lahul, Spiti and

Red-fronted Serin, photograph by Mubashir Hasan

Garhwal between elevations of 2,400 and 4,000 metres. During winter, it descends to lower altitudes frequenting open hillsides and stony ground with bushes (1.10:147).

In non-breeding season, it keeps in restless flocks, feeding mostly on the ground on seeds, grasses and weeds. It is an avid water drinker and may be seen bathing and drinking at all times of the day.

The Red-fronted Finch is smaller in size than the sparrow, and has a bright scarlet-orange forehead, a dark brown crown and blackish back of the neck. The back is yellowish buff heavily streaked with dark brown. The throat, the sides of the head and the breast are brown. The belly and flanks are pale yellow with dark brown streaks. Both sexes look alike.

Body length 120 mm.

Himalayan Greenfinch
Carduelis spinoides or *Carduelis chloris*

The Greenfinch is found in a relatively narrow north-south strip of the Karakorams and the Himalayas extending from Swat to Assam. It is also found in Burma, Vietnam and south-west China. It frequents open mountain slopes, cultivations, orchards, gardens and the edges of forests. It feeds much more in the bushes and flower-heads than on the ground and is easily attracted to gardens by planting sunflowers. It breeds between June and October at elevations ranging between 1,300 and 4,400 metres and descends to the foothills and plains in October. It is a common winter visitor to the Peshawar valley (4:222).

The upper plumage of the Greenfinch is greenish-brown mixed with black; the wings are dark brown having markings of yellow, black and a little white. The rump and the whole of the lower plumage is bright yellow. The female is like the male but the colours are paler.

Body length 140 mm.

Himalayan Greenfinch, photograph by Syed Asad Ali

Goldfinch
Carduelis carduelis
Shaira or *Sehara* in Kashmiri

The Goldfinch is found in northern Balochistan, the NWFP, Chitral, Kaghan, Gilgit, Baltistan, Kashmir and east up to central Nepal; also found throughout Europe and Central Asia. It frequents orchards, pine and fir forests and bare stony hillsides. It feeds on grass seeds often perched on flower-heads and is reported to breed between elevations of 1,500 to 3,300 metres. It has a long tweezer-like bill for probing into thistles and other seed-heads. It descends to lower altitudes in winter and keeps in pairs and small flocks.

Smaller in size than the Sparrow, the Goldfinch is a beautifully coloured bird with a crimson forehead, chin and area around the bill; black wings with bright yellow patch; whitish rump and belly. The crown, back and breast are pale grey-brown. In the subcontinent, it lacks the black head that its race wears in Europe. Both sexes look alike.

Body length 140 mm.

Goldfinch, photograph by Mubashir Hasan

Trumpeter Bullfinch, photograph by Mark Malallieu

Trumpeter Bullfinch or Trumpeter Finch
Rhodopechys githaginea or *Bucanetes githagineus*

The Trumpeter Bullfinch is a smaller than sparrow-sized bird of arid areas. It is found in Afghanistan, parts of Central Asia, Iran, Iraq and west up to Morocco. In Pakistan it is found mostly to the west of the river Indus, in Balochistan and the NWFP (2.2:538) and in similar desert conditions in Rajasthan (Roberts' personal comments). It frequents stony slopes and steep ravines and feeds on grass and its flocks fly out regularly in the morning and evening to drink at desert springs.

The Trumpeter Bullfinch in breeding plumage is a pale pinkish-brown finch with a heavy orange bill. Its back is greyish-brown washed with pink; the crown is ashy-grey; the wings and lower rump are pink. The lower plumage is pale greyish-pink. The female is similar to the male but without the pink tinge during winter.

Body length 150 mm.

Common Rosefinch or Scarlet Grosbeak
Carpodacus erythrinus
Tuti, Lal Tuti, Gulabi Tuti in Urdu, Hindi and Bengali
Gulab Tsar in Kashmiri

The Rosefinch is a sparrow-sized bird which breeds in the hills of Pakistan and India from northern Balochistan, Chitral, Kaghan, Gilgit, Baltistan, Kashmir, Ladakh, Lahul and Spiti, east up to Arunachal Pradesh and beyond. During winter it descends to the foothills and spreads all over the subcontinent except Sindh and southern Balochistan. It is also found in Afghanistan, parts of Iran, Central Asia and eastern Europe. It feeds on seeds, flower buds, nectar, berries and fruit.

The entire body plumage of the male Rosefinch is dull crimson mixed with brown on the back and sides. The crimson is brighter on the rump, chin, throat and breast; the wings and tail are brown edged with rufous. The entire plumage of the female is olive brown, streaked with brown.

Body length 150 mm.

Common Rosefinch, photograph by Mubashir Hasan

BUNTINGS

Family Emberizinae

The members of the family *Emberizinae* are sparrow-sized, ground-living birds or dwell in bushes and thickets. Buntings can be found anywhere in Asia and Africa, in all sorts of climates but mostly in temperate regions. Their plumage is brown, grey or olive with combinations of black, white, yellow, green or red. They mainly eat seeds and have short, conical and finely pointed bills adapted for the purpose. During breeding season many buntings rear their young largely or entirely on arthropod prey (19:73). Their flight is fairly fast and undulating.

White-capped Bunting
Emberiza stewarti

The White-capped Bunting is not a common bird but where found, this sparrow-sized bird is quite numerous. It breeds in northern Balochistan, the valleys of Kurram, Chitral, Gilgit, Baltistan, Kashmir, Kullu, Kangra, and Garhwal between elevations of 1,200 and 3,300 metres. It also breeds in Afghanistan, eastern Iran and parts of Central Asia. During winter it spreads over parts of central Punjab, western NWFP in Pakistan and over the Uttar Pradesh, Rajasthan and Kathiawar in India. It is generally found in small flocks, which feed mainly on seeds of grass.

The top of the head of the male White-capped Bunting is grey; the back and rump are chestnut; the breast is white and

White-capped Bunting, photograph by Mubashir Hasan

the lower breast chestnut. The upper plumage of the female is ashy-brown streaked with blackish except on the sides of the face; the wings are brown and the lower plumage is pale fulvous streaked with brown.

Body length 150 mm.

Rock Bunting
Emberiza cia
Wantsar in Kashmiri

The Rock Bunting is a sparrow-sized bird with a striking head pattern. It breeds from May to August in the higher hills of the NWFP, the valleys of Gomal, Kurram, Chitral, Kaghan, Gilgit, Baltistan, Kashmir, Lahul and Spiti between elevations of 2,700 and 3,300 metres. It also breeds in southern Europe, Turkey, Iran, Afghanistan and parts of Central Asia. During winter it spreads over most of the Punjab and the NWFP and over northern India. It feeds on seeds of grass, grains and insects.

Rock Bunting, photograph by Mubashir Hasan

The upper plumage of the Rock Bunting is chestnut brown; streaked darker, the head, throat and upper breast are pale bluish-grey, marked with two black lines along the top of the head. The rest of the lower plumage is rufous chestnut. Both sexes look alike.

Body length 150 mm.

Striolated Bunting, House Bunting or Striped Bunting
Emberiza striolata

The Striolated Bunting is found in Balochistan, Sindh and some areas of the Punjab. It is also found in North Africa, the Arabian Peninsula, Iraq, and western parts of central India (1.10:229). It is as common and familiar a bird in villages and cities of North Africa as the House Sparrow is in this part of the world. It feeds on seeds of grasses and frequents sandy plains, tamarisk scrub, rocky, sparsely scrubbed hillsides and nullahs. In arid country it regularly drinks at waterholes in the forenoon.

Striolated Bunting, photograph by Syed Asad Ali

The upper plumage of the male Striolated Bunting is brown with darker streaks. The crown, cheeks and throat are greyish-white streaked with blackish, the cheeks are bordered with black. The belly is fulvous buff. The female is similar to the male but its head and throat are brown with darker streaks.

Body length 140 mm.

Crested Bunting
Melophus lathami
Patthar Chirya in Urdu and Hindi

The Crested Bunting, a sparrow-sized bird, is found in the foothills of the Himalayas between elevations of 1,500 and 2,400 metres, from Hazara in Pakistan to Bhutan in the east and in central parts of the Indian Union. It feeds on the ground, mainly on grass seeds and, in winter, frequents sparsely scrubbed hillsides and rice stubbles. It is particularly fond of charred grass.

Crested Bunting, photograph by Mubashir Hasan

The entire body plumage of the Crested Bunting is black. The wings and tail are bright chestnut and it has a long pointed crest. The female is olive brown with dark brown streaks. Its crest is shorter and the lower plumage is yellowish buff with dark streaks on the breast.

Body length 150 mm.

REFERENCES*

1. Ali, S. and Ripley S.D. 1971–1981. *Handbook of the Birds of India and Pakistan*. 10 Volumes. Oxford University Press.

2. Roberts, T.J. 1991–1992. *The Birds of Pakistan*. Oxford University Press.

3. Ali, Salim. 1972. *The Book of Indian Birds*. Bombay Natural History Society.

4. Whistler, Hugh. 1963. *Popular Handbook of Indian Birds*. Oliver and Boyds Ltd.

5. Heinzel, Fitter, Parslow. 1984. *The Birds of Britain and Europe*. Collins.

6. Ginn, P.J., McIlleron, W.G., Milstein, P. le S. 1989. *The Complete Book of Southern African Birds*. Struik Winchester.

7. Welty, Joel Carl. 1982. *The Life of Birds*. Saunders College Publishing.

8. Bull, John and Farrand, John Jr. 1977. *The Audubon Society Field Guide to North American Birds*.

9. Woodcock, Martin. *Collins Handguide of the Birds of the Indian Subcontinent*.

10. Ripley, Sidney Dillon, 1982. *A Synopsis of the Birds of India and Pakistan together with those of Nepal, Bangladesh and Sri Lanka*. BNHS.

---

* Guide to references:
(2.1:182) = the publication at serial no. 2 in the list of 'References'; volume; and page number
(7:594) = the publication; page number

11. *A New Dictionary of Birds*, 1974. Edited by A. Lansborough Thomson. Centenary Publication of the British Ornothologist Union.

12. Boonsong Lekagul and Pilip D. Round, 1991. *A Guide to the Birds of Thailand*. Saka Karn Bheet Co. Ltd.

13. Law, Satya Churn, 1923. *Pet Birds of Bengal*. Thacher Spink & Co.

14. *Handbook of the Birds of Europe, the Middle East and North Africa, The Birds of the Western Palearctic, 1977*. Volume 1. Stanley Cramp, Chief Editor. Oxford University Press.

15. As above. Volume 2. *Hawks to Bustards*. 1980.

16. As above. Volume 3. *Waders to Gulls* 1983.

17. As above. Volume 4. *Terns to Woodpeckers*. 1985.

18. Baker, E.C. Stuart, 1913. *Indian Pigeons and Doves*. Witherby & Co.

19. *A Dictionary of Birds*, 1985. Editors, Bruce Campbell and Elizabeth Lack. Calton (Poyser) and Vermillion (Buteo).

20. Austin, Oliver L. Jr. 1983. *Birds of the World*. Optimum Books.

21. Macdonald, Malcolm, 1960. *Birds in My Indian Garden*. Jonathan Cape.

22. Hume and Marshall, 1880. *The Game Birds of India, Burmah and Ceylon*. Volumes 1–3. Published by the authors.

23. Baker, E.C. Stuart, 1922–30. *Fauna of British India: Birds*. 8 Volumes. Taylor and Francis.

24. Asian Water fowl Census, 1991, International Water fowl and Wetlands Research Bureau. Slumbridge, Gloucester. U.K.

25. Dorst, Jean, 1962. *The Migration of Birds*. Heinemann.

26. Mc Cure, H. Elliott, 1974. *Migration and Survival of the Birds of Asia*. U.S. Army Medical Component, South East Asia Treaty Organisation, Bangkok, Thailan.

27. Journal of Bombay Natural History Society. Volume 47.

28. Ganguli, Usha, 1975. *A Guide to the Birds of Delhi Area.* Indian Council of Agricultural Research.

29. Wood-Gush, D.G.M. 1964 'Domestication' in Thompson, A.L. (Ed.). *A New Dictionary of Birds.* Mc Graw Hill.

INDEX OF ENGLISH NAMES

INDEX OF SCIENTIFIC NAMES

INDEX OF URDU/HINDI NAMES